Harwood Fundamentals of Pure and Applied Economics

THE ROLE OF DEMAND AND SUPPLY IN THE GENERATION AND DIFFUSION OF TECHNOLOGICAL CHANGE

Routledge
Taylor & Francis Group
LONDON AND NEW YORK

ECONOMICS OF TECHNOLOGICAL CHANGE II
In 2 Volumes

THE ROLE OF DEMAND AND SUPPLY IN THE GENERATION AND DIFFUSION OF TECHNICAL CHANGE

COLIN G THIRTLE AND
VERNON W RUTTAN

Routledge
Taylor & Francis Group
LONDON AND NEW YORK

First published in 1987 by
Harwood Academic Publishers GmbH

Reprinted in 2001 by
Routledge

2 Park Square, Milton Park, Abingdon, Oxfordshire OX14 4RN

711 Third Avenue, New York, NY 10017

First issued in paperback 2014

Routledge is an imprint of the Taylor & Francis Group, an informa business

British Library Cataloguing in Publication Data
A CIP catalogue record for this book
is available from the British Library

The Role of Demand and Supply in the Generation and Diffusion of Technical Change

ISBN 978-0-415-26934-6 (hbk)
ISBN 978-0-415-75369-2 (pbk)

Economics of Technological Change II: 2 Volumes
ISBN 978-0-415-26932-2
Harwood Fundamentals of Pure & Applied Economics
ISBN 978-0-415-26907-0

The Role of Demand and Supply in the Generation and Diffusion of Technical Change

Colin G. Thirtle
University of Manchester, UK

and

Vernon W. Ruttan
University of Minnesota, USA

A volume in the Economics and Technological Change section

edited by

F. M. Scherer
Swarthmore College, USA

harwood academic publishers
chur · london · paris · new york

Harwood Academic Publishers

Post Office Box 197
London WC2E 9PX
England

58, rue Lhomond
75005 Paris
France

Post Office Box 786
Cooper Station
New York, NY 10276
United States of America

Library of Congress Cataloging-in-Publication Data

Thirtle, Colin G., 1944–
The role of demand and supply in the generation and diffusion of technical change.

(Fundamentals of pure and applied economics,
ISSN 0191-1708; vol. 21. Economics of technological change section).
Bibliography: p.
Includes index.
1. Technological innovations—Economic aspects.
2. Supply and demand. I. Ruttan, Vernon W. II. Title.
III. Series: Fundamentals of pure and applied
economics; vol. 21. IV. Series: Fundamentals of pure
and applied economics. Economics of technological
change section.
HC79.T4T495 1987 338'.06 87-138
ISBN 3-7186-0384-5

Contents

Introduction to the Series

Drawing on a personal network, an economist can still relatively easily stay well informed in the narrow field in which he works, but to keep up with the development of economics as a whole is a much more formidable challenge. Economists are confronted with difficulties associated with the rapid development of their discipline. There is a risk of "balkanisation" in economics, which may not be favorable to its development.

Fundamentals of Pure and Applied Economics has been created to meet this problem. The discipline of economics has been subdivided into sections (listed inside). These sections include short books, each surveying the state of the art in a given area.

Each book starts with the basic elements and goes as far as the most advanced results. Each should be useful to professors needing material for lectures, to graduate students looking for a global view of a particular subject, to professional economists wishing to keep up with the development of their science, and to researchers seeking convenient information on questions that incidentally appear in their work.

Each book is thus a presentation of the state of the art in a particular field rather than a step-by-step analysis of the development of the literature. Each is a high-level presentation but accessible to anyone with a solid background in economics, whether engaged in business, government, international organizations, teaching, or research in related fields.

Three aspects of *Fundamentals of Pure and Applied Economics* should be emphasized:

—First, the project covers the whole field of economics, not only theoretical or mathematical economics.

—Second, the project is open-ended and the number of books is not predetermined. If new interesting areas appear, they will generate additional books.
—Last, all the books making up each section will later be grouped to constitute one or several volumes of an Encyclopedia of Economics.

The editors of the sections are outstanding economists who have selected as authors for the series some of the finest specialists in the world.

J. Lesourne *H. Sonnenschein*

The Role of Demand and Supply in the Generation and Diffusion of Technical Change

COLIN G. THIRTLE and VERNON W. RUTTAN

University of Manchester and University of Minnesota

INTRODUCTION

The importance of both science and technology for modern economic growth has been accepted as almost self-evident since at least the middle of the nineteenth century [283, p. 355; 342]. But it was not until the mid-1950s that economists attempted to measure the contribution of technical change to economic growth [1, 444, 480, 506].

The primary focus of the early studies on technical change and productivity growth was simply to measure the contribution of technical change, relative to conventional resources, to growth in output. Technical change itself was treated as a response to the economic opportunities resulting from autonomous advances in scientific and technical knowledge.[1] By the mid-1960s, however, increasingly serous efforts were being made to explore the influence of economic forces on technical change.

In this paper we attempt to review and assess the literature on the

---

Colin G. Thirtle is Lecturer in the Department of Agricultural Economics, University of Manchester. Vernon W. Ruttan is Regents Professor in the Department of Agricultural and Applied Economics and in the Department of Economics, University of Minnesota.

The authors are indebted to John Chipman, Glen Fox, Zvi Griliches, Yujiro Hayami, Onko Kingma, J. S. Metcalfe, Richard R. Nelson, Willis Peterson, Nathan Rosenburg and F. M. Scherer for comments on an earlier draft of this paper.

[1] Rosenberg notes, "This view has the disconcerting aspect, at least for the economist, of appearing to make the central feature of modern economic growth an exogenous phenomenon . . . . Economists have had much more success in dealing with the *consequences* of technological change than with its determinants" [434, p. 141].

impact of economic forces on the rate and direction of technical change. The paper begins by considering the impact of economic forces on invention and innovation. In Part 2 we examine the impact of factor endowments and prices on the direction of technical change, and in Part 3 we examine the process of the diffusion of technology.

1. SUPPLY AND DEMAND EXPLANATIONS OF INVENTION AND INNOVATION

Schumpter, whose writings have been exceptionally important in formulating the way economists think about technical change, made a sharp distinction between invention (and the inventor) and innovation (and the innovator): "Innovation is possible without anything we should identify as invention, and invention does not necessarily induce innovation but produces itself . . . no economically relevant effect at all" [487, Vol. 1, p. 84]. Rosenberg has argued that the effect has been to divert the attention of economists away from those activities that are most relevant to technical innovation and the diffusion or transfer of technology [432, pp. 66–68]. Other students of technical change have argued that the Schumpeterian distinction between invention and innovation is excessively artificial. For analytical purposes it is more useful to use the term *innovation* to designate any "new thing" in the area of science or technology and to reserve the term *invention* to refer to that subset of technical innovations that are patentable [263, p. 2; 351, p. 103; 445, p. 605].

1.1. Processes of invention

At the time economists first became interested in the economics of invention and innovation there were already well-defined traditions of scholarship in the literature on applied technology, sociology, and history. In his classic study, *A History of Mechanical Inventions,* Usher [547, pp. 56–83] identified three general approaches to the emergence of inventions. He termed these the *transcendentalist* approach, the *mechanistic process* approach, and the *cumulative synthesis* approach.

The *transcendentalist approach* attributes the emergence of invention to the inspiration of the occasional genius who from time to time achieves insight into essential truth through the exercise of personal energy, intuition, and skill. This heroic approach to the process of invention bears striking resemblance to the Schumpeterian view of the entrepreneur. The transcendentalist perspective dominated much of the early historical and biographical scholarship on technical change [234, 501].[2] Usher rejected the transcendentalist view as unhistorical. He argued that the act of insight was not the rare, unusual phenomenon assumed by the transcendentalists and further that the act of insight that results in the perception of new relationships requires a highly specific conditioning of the mind within the framework of the problem to be solved. It was not an accident that Henry Ford, a bicycle mechanic, contributed to the development of the automobile or that Harry Ferguson, a self-taught mechanic, was the first to apply basic physical principles to the integrated design of tractors and tractor equipment.

The *mechanistic process* theory viewed invention as proceeding under the stress of necessity with the individual inventor being an instrument of historical processes. This view emerged from the detailed investigations of invention sequences by the Chicago sociologists, Ogburn and Gilfillan.[3] By demonstrating that the process of invention typically represented a new combination of a large number of individual elements accumulated over long periods of time, the sociologists erected an effective challenge to the claims of the transcendentalists. But Usher argued that the approach overlooked the significance of discontinuities inherent in the process of invention and insisted that the "acts of insight" required to bridge the discontinuities are possible for only a limited number of individuals operating under conditions that bring both an awareness of the problem and the elements of a solution within their frame of reference. And even under these conditions it is not certain that the specific act of insight required for a solution to the problem will occur.

---

[2] Although Smiles' portrait of the character and accomplishments of the early British engineers was cast in the heroic mode, he was sensitive to the political, economic, and social forces that influenced their accomplishments [234, p. 5].

[3] Both Ogburn and Gilfillan emphasized that inventions generally occur incrementally as a result of the accumulation of experience, rather than as dramatic breakthroughs. Ogburn, however, placed primary emphasis on the rate of advances in knowledge and on the state of "material culture" while Gilfillan placed greater emphasis on the response to demand [169, 170, 171, 375, pp. 30–102; 376, pp. 775–810; 377].

Usher suggested a *cumulative synthesis* approach as an alternative to the transcendentalist and mechanistic process theories of invention. With this framework, which drew on Gestalt psychology, major inventions are visualized as emerging from the cumulative synthesis of relatively simple inventions, each of which requires an individual "act of insight." A major or strategic invention, or advance in technology, represents the cumulative synthesis of many individual inventions. Many of the individual inventions do no more than set the stage for a major invention that then requires substantial critical revision to adapt it to a particular use. A schematic presentation of the elements of the individual act of insight and the cumulative synthesis as visualized by Usher are presented in Figures 1.1 and 1.2.

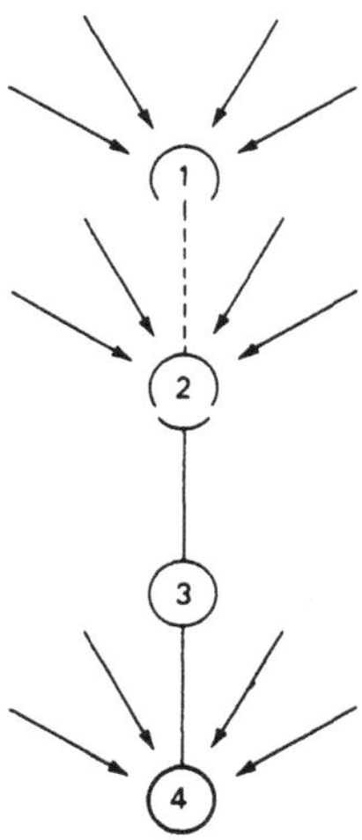

FIGURE 1.1 The emergence of novelty in the act of insight. Synthesis of familiar items: (1) perception of an incomplete pattern; (2) setting the stage; (3) the act of insight; and (4) critical revision and full mastery of the new pattern. (After Abbott P. Usher, *A History of Mechanical Inventions* [Cambridge, Mass.: Harvard University Press, 1954].)

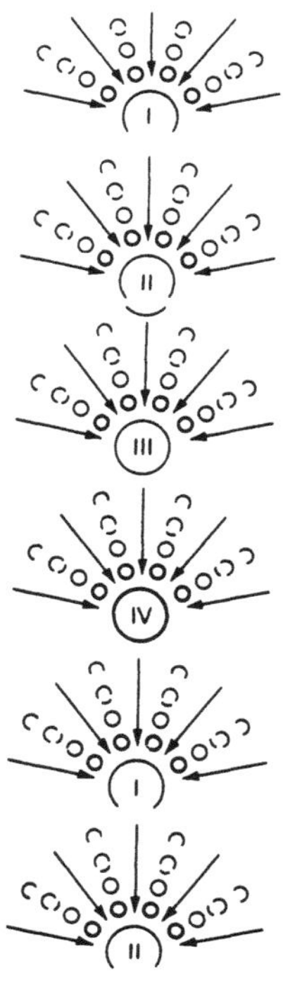

FIGURE 1.2 The process of cumulative synthesis. A full cycle of strategic invention and part of a second cycle. Roman numerals I through IV represent steps in the development of a strategic invention. Small figures represent individual elements of novelty. Arrows represent familiar elements included in the new synthesis. (After Abbott P. Usher, *A History of Mechanical Inventions* [Cambridge, Mass.: Harvard University Press, 1954].)

Usher's cumulative synthesis approach provides the element of a critical theory of the social process by which "new things" come into existence and are improved, a process that is broad enough to encompass the whole range of activities characterized by the terms *science, invention,* and *innovation.* One is no longer forced to maintain, as Schumpeter did, the increasingly artificial distinction between the processes of invention and innovation or to explain away the association between scientists, inventors, and entrepreneurs as merely a chance coincidence. But the Schumpeterian system has remained an obstacle to the efforts by economists to understand the processes of technical innovation [432, pp. 66–68].

A major contribution of Usher's cumulative synthesis theory was that it clarified the points at which economic forces could be used to speed the rate or alter the direction of technical change. The possibility of allocating research resources to influence the rate or direction of technical change was obscured by the transcendentalist approach, with its dependence on the emergence of the hero inventor, and was denied by the mechanistic process approach, with its dependence on inexorable historical trends or forces. The focus of conscious effort to affect the rate or direction of technical innovation centers around the second and fourth steps in the process as outlined by Usher—in setting the stage and in critical revision. By consciously bringing together the elements of a solution—by creating a favorable environment—the stage can be set to enhance the probability that the critical act of insight will occur. Two of the great institutional innovations of the nineteenth century, the industrial research laboratory and the agricultural experiment station, were consciously designed to set the stage more effectively for technical innovation.[4]

The impulses that gave rise to both the transcendentalist and mechanistic process approaches continue to be reflected in contemporary efforts to understand the forces that influence the rate and direction of technical change. The transcendentalist perspective was essentially a supply-side perspective. Its equivalent is the contemporary view that autonomous advances in scientific and technical knowledge determine the rate and direction of technical change.

[4] Alfred North Whitehead has argued that "the great invention of the nineteenth century was the invention of the method of invention" [562, p. 96].

There are both supply-side and demand-side variants of the mechanistic process perspective. In the supply-side variant, technical change is a near automatic response to advances in material culture or knowledge, and in the demand-side variant it is a near automatic response to growth in product demand or to changes in relative factor prices.

The dialogue about the sources of technical change continues to center around whether technical change has been driven primarily by autonomous advances in science and technology or driven primarily by economic forces—whether technical change is most appropriately viewed as exogenous or endogenous to the economic system. In the next section we review the recent dialogue and evidence on this issue.

1.2. Sources of technical change: demand pull and supply push

Before the beginning of the nineteenth century the linkages between advances in scientific knowledge and advances in technology were relatively weak. "Science was traditionally aristocratic, speculative, intellectual in intent; technology was lower-class, empirical and action oriented" [561, p. 79]. Science had remarkably little to offer to those who were engaged in advancing technology.

The nineteenth century witnessed a remarkable fusion of theoretical and empirical inquiry [287]. By the middle of the twentieth century a new orthodoxy had emerged to the effect that modern technology was simply applied science [27]. Basic science developed theory and understanding; applied science took that knowledge and used it in the design of new technology. In the United States this new orthodoxy was reinforced by the success of World War II science-based military technology. It found its most influential expression in the report by Vannevar Bush on post-war scientific research [77]. The Bush report became the charter for post-war science and technology policy. And the science-based technology development perspective tended to dominate many of the early post-war studies of invention and innovation [302].

The interaction between advances in science and technology is, however, much more complex than is reflected in the early post-war perspective [168, 287, 400]. Instead of a single path running from scientific discovery through applied science to development, it is

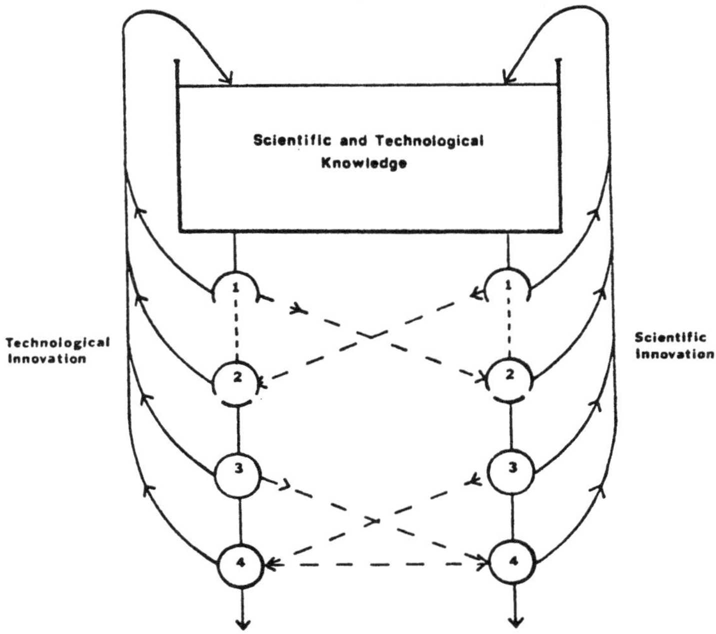

FIGURE 1.3 Technological and scientific innovation

more consistent with historical evidence to model science-oriented and technology-oriented research as two parallel but interacting paths. These two paths are connected through a common pool of existing scientific knowledge; both paths lead from and feed back into further advances in both scientific and technical knowledge (Figure 1.3).

Since the early 1960s, an increasingly serious challenge has been mounted against the new orthodoxy. The challenge has proceeded along three fronts. One has been to challenge the historical accuracy of the view that the flow of knowledge has run in a linear sequence from science to technology.[5] A second has been to

[5] After an extensive review of the history of a number of major innovations, Rosenberg asserts that "the normal situation in the past, and to a considerable degree also in the present, is that technological knowledge had *preceded* scientific knowledge . . . . It is still far from unusual for engineers in many industries to solve problems for which there is no scientific explanation, and for the engineering solution to *generate* the subsequent scientific research that eventually provides the explanation" [434, p. 144].

document that the allocation of resources to inventive activity and to research has been strongly influenced by changes or differences in demand.[6] A third has been an attempt to show that the development and diffusion of commercially successful technical innovations have been primarily a response to changes (or differences) in demand.[7]

Arguments about the priority of the role of market demand and the supply of knowledge in inducing advances in technology were intensified by the late 1960s by a study conducted by the Office of the Director of Defense Research and Engineering (HINDSIGHT) [374, 497] that purported to show that the significant "research events" that had contributed to the development of 20 major weapons systems were predominantly motivated by military need rather than disinterested scientific inquiry. This view was challenged in studies commissioned by the National Science Foundation and conducted by the Illinois Institute of Technology (TRACES) [236] and the Battelle Research Institute [30]. The TRACES and Battelle studies adopted a much longer time horizon than the 20-year period employed in the HINDSIGHT study. And, not unexpectedly, they found that science events were of much greater importance, relative to technology events, as a source of technical change than was shown in the HINDSIGHT study.

From an analytical perspective, the "demand pull" or "demand-induced" theory of technical change can be thought of as a scheme in which the demand for technical change, in the form of product and process innovations, is derived from the demand for commodities; the demand for inventive activity, including research and development, is derived from the demand for technical change; and the demand for advances in scientific knowledge is, in turn, derived from the demand for inventive activity.

In spite of the large literature on the influence of market demand on technology development, the evidence on the relative significance of "demand pull" on the rate and direction of technical change has not been firmly established. Mowery and Rosenburg

[6] The proposition that the rate of invention in the capital goods industries in the United States is closely associated with the rate of capital investment was established by Schmookler [481, pp. 104–163] and confirmed by Scherer [478].

[7] This literature has been critically reviewed by Mowery and Rosenberg [346, 434]. See also the review by Kamien and Schwartz [263, pp. 31–47].

argue that much of the recent research purporting to show that technical innovation has largely been demand-induced is seriously flawed by lack of rigor in the specification of demand [346, 434, pp. 192–241].[8] The demand pull model of technological change has also been criticized for ignoring both the internal logic of scientific progress and the historical contribution of science to technical progress. Much of the earlier research in the philosophy and history of science and in the sociology of knowledge presumes that advances in science are largely determined by the internal logic of discovery in the several scientific disciplines [280, 400].[9] The demand pull perspective has been criticized as ignoring "the whole thrust of modern science and the manner in which the growth of specialized knowledge has shaped and enlarged man's technological capacities" [431, 432, p. 264].

The supply or technology push view, restated in economic terms, is that autonomous advances in scientific and technical knowledge permit the substitution of calculation or computation for the more expensive process of trial and error. The effect of advances in science is to shift the supply curve for technical change to the right [351, p. 106]. A classical example is the invention of the contact process for sulfuric acid. Mathematical modeling indicated with great precision that there was only one practical way to achieve synthesis of sulfuric acid by the contact process [351, pp. 105-111].

Rosenberg has suggested that "to establish the independent importance of supply-side considerations, it is necessary to demonstrate several things: (1) That science and technology progress, in some measure, along lines determined either by internal logic, degree of complexity or at least in response to forces independent of economic need; (2) that this sequence in turn imposes contraints

[8] Rosenberg and Mowery argue that much of the research that purports to demonstrate the plausibility of the demand pull hypothesis can more appropriately be interpreted as evidence that private profitability or social utility (or need) are inducements to the allocation of private and public resources to research. But profitability and utility are also enhanced by advances in scientific or technical knowledge that reduces the cost of technical change. See also Nelson [351].

[9] This view has also been characteristic of much of the anti-technology movement literature. For an extreme view of the autonomy of science see Mishan: ". . . science is not guided by any social purpose . . . . As a collective enterprise science has no more social conscience than the problem solving computers it employs. Indeed, like some ponderous multipurpose robot that is powered by its own insatiable curiosity, science lurches onward . . ." [339, p. 129].

or presents opportunities which materially shape the direction and the timing of the inventive process; and (3) that, as a result, the costs of invention differ in different industries" [432, pp. 265–266].

Our review of the literature does not lead us to a rejection of the "demand pull" model of technical change in spite of the Mowery–Rosenburg criticisms. The model is supported by rigorous studies at both the sector [190, 481] and macroeconomic levels [36, pp. 261–275; 301]. The model has also provided the theoretical framework for the estimates of the rates of return to research developed initially by Griliches [191].

It is also our view that the dialogue over the relative priority of "demand pull" and "supply push" explanations of the rate and direction of technical change has been misplaced. It is not necessary to demonstrate that basic research is the cornucopia from which all inventive activity must flow to conclude that investment in the generation of new scientific and technical knowledge can open up new possibilities for technical change. Nor is it necessary to demonstrate that advances in knowledge, inventive activity, and technical change flow automatically from changes in demand to conclude that changes in demand represent a powerful inducement for the allocation of resources to research [351].

What can be said at this stage about the relative importance of the demand pull or demand-induced theory, and of the supply or technology push theory, in accounting for the rate and direction of technical change? The only study we have been able to identify that attempts to test simultaneously the demand-induced and the supply push hypotheses was conducted by Scherer [478]. The Scherer analysis confirmed the earlier Schmookler findings of strong association between capital goods inventions and investment. But Scherer found that the association between industrial materials inventions and measures of demand pull (materials purchased and value added) was considerably weaker than in the capital goods industries. He also found that introduction of an index of technological opportunity, based on the richness of an industry's knowledge base, added significantly to the power of his model to explain differences in the level of invention activity among industries.

Both private and public sector research managers are faced with questions about the relative priority of the allocation of research resources (a) to advance knowledge in those fields in which scientific and technological opportunities appear most favorable or (b) for applied research and development in those industries characterized by current or anticipated rapid growth in demand. The research we have reviewed in this section gives us little guidance beyond Nelson's conclusions of a quarter century ago: "Though the expected profitability of an invention in a particular field affects the rate of invention activity in that field, the tremendous uncertainties involved in making any major technological breakthrough preclude either the routinization of invention or the precise prediction of invention. Conditions of demand and of scientific knowledge provide us with guides for prediction and analysis, but only with rough guides" [351, p. 115].

The dialogue with respect to the role of "demand pull" and "supply push" has been relatively unproductive in the generation of either a rigorously testable empirical proposition or a useful guide to research policy. A second body of literature that has focused on the impact of changes or differences in resource endownment and factor prices on the direction of technical change has been much more productive. We turn in the next part to a review of the literature on factor bias and induced technical change.

2. INDUCED INNOVATION AND FACTOR BIASES

The total factor productivity studies[10] of the 1950s showed that only a small proportion of the long-term growth of output of the American economy could be explained by conventionally measured increases in the quantities of labour and capital. Solow's [506] classic estimation, consistent with his own growth model [505], attributed 87.5% of per capita growth to technical change and 12.5% to increased capital per head.

The unexpected importance of the residual measure of technical change or productivity growth[11] led to a revival of interest in the classification and explanation of changes in technology.

[10] This area has been reviewed by Nadiri [349]. More recent developments are discussed in [268] and [121]. Few productivity studies are referred to in this paper since the typical objective of such work is a total factor productivity index that describes the rate but not the *bias* of technical change.

[11] Productivity growth and technical change were assumed to be synonymous, but productivity is also affected by increasing returns to scale and improvements in the allocation of resources, such as structural change [508, p. 93].

2.1. Neutrality and bias in technical change

The term *technical progress* has been used to describe both increases in the stock of knowledge pertaining to the art of production and the effects of such new technology on the level of output [272]. The increase in output obtained from the same quantities of inputs, or equivalently,[12] the decrease in inputs required to produce a given level of output, provides a measure of technical progress.[13] This definition of technical change is appealing because it is straightforward and consonant with the neoclassical representation of technical change as the movement of the isoquant towards the origin (Figure 2.1), or an upward shift of the production function.[14] The disadvantage is that technical change is unlikely to affect all factor inputs equally. This difficulty, which raises the issue of factor bias in technical change, has led several authors to define technical change in terms of the proportional decrease in production costs at constant factor prices.[15]

2.1.1. Hicks neutrality and the microeconomic approach to technical change. In Figure 2.1 inputs of labor and capital are measured in homogenous physical units and all four isoquants represent a fixed level of homogenous output. Initial equilibrium is at point A where the isoquant I_0 is tangential to the isocost constraint PP'. Since the slope of the isoquant is determined by the ratio of the marginal products of the two inputs and the slope of PP' represents the ratio of factor prices, the two ratios must be equal at equilibrium points such as A. The isoquants I_1, I_2 and I_3 represent alternative new

---

[12] If the production function is homogeneous of degree one (constant returns to scale), it can be represented by any single isoquant and the two measures will be identical.

[13] This measure is suited to process innovations, which may be thought of as better ways of making existing goods. Product innovation, encompassing both the appearance of new goods and improvements in quality, while very important [283, pp. 339–346], has proved less tractable. Thus, theories of induced innovation are confined to process innovation. This serious limitation is mitigated by the fact that the product innovations of the capital and intermediate goods industries can be viewed as process innovations in the industries using their output [61].

[14] The production function is taken to embody "all previously known techniques" [272].

[15] This definition is suitable for micro studies since the individual firm can treat prices as exogenous. If the firm is using the cost minimizing input combination before and after the change, both situations will lie on the expansion path.

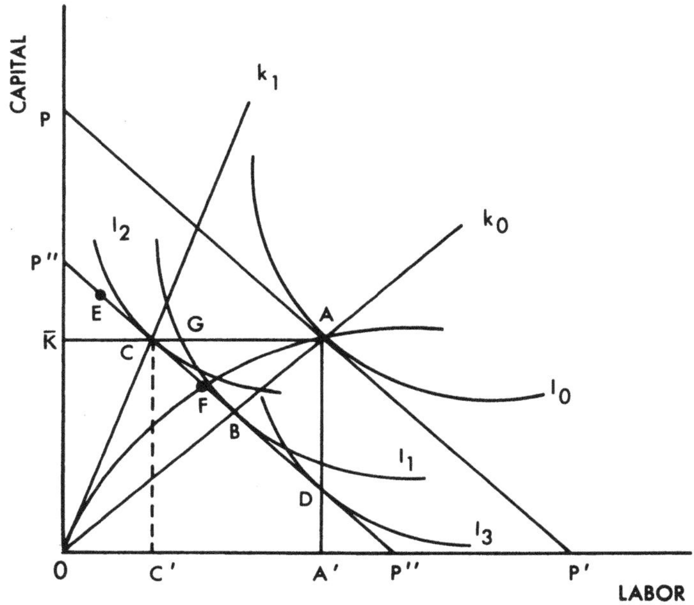

FIGURE 2.1 Neutrality and bias of technical change at constant factor prices.

technologies. For all three, the proportional saving in resources is *PP″/OP*, but the factor-saving biases differ. If the new technology results in a new equilibrium at point *B* on isoquant I_1, then technical change saves both factors in the same proportion as they were being used and the original factor ratio k_0 is maintained. This is the notion of neutrality in technical change associated with Hicks [225].

If the production function can be written in the factor-augmenting form,

$$Y = F[A(t)K, B(t)L], \tag{1}$$

Hicks neutrality requires that

$$\frac{\dot{A}(t)}{A(t)} = \frac{\dot{B}(t)}{B(t)} = m, \tag{2}$$

so that technical change augments[16] both factors equally, and if constant returns to scale are assumed, the production function can be written as

$$Y = A(t)F(K, L). \tag{3}$$

Bias is defined relative to neutrality, so that if the new equilibrium is at any point to the northwest of B, along $P''P''$, technical change is said to be labor-saving. Since the proportional reduction in labor is greater than that for capital, the result is an increase in the capital-labor ratio. For example, if the new equilibrium were at C on isoquant I_2, the proportional reduction in labour input is $C'A'/OA'$, while the input of capital remains unchanged and the capital–labor ratio is increased to k_1. In terms of the factor-augmenting production function of Eq. (1), this requires that

$$\frac{\dot{A}(t)}{A(t)} = 0 \quad \frac{\dot{B}(t)}{B(t)} = m_l. \tag{4}$$

Technical change is *purely* labor-saving or augmenting, but this is not a limiting case. A point such as E could also represent the new equilibrium, in which case the change is labor-saving and capital-using in the sense that absolutely more capital is required. Unfortunately the term *capital-using* has been applied to all changes that save a larger proportion of labor (relative to initial usage) than of capital (all points to the northwest of B). Similarly, point D represents a purely capital-saving technical change, which in terms of factor augmentation requires that

$$\frac{\dot{A}(t)}{A(t)} = m_k, \quad \frac{\dot{B}(t)}{B(t)} = 0. \tag{5}$$

To summarize, Hicks neutrality and bias can be defined [53, 462] in terms of the proportional change in the capital–labor ratio at

[16] Augmentation is a particular representation of technical change in the production function, not an explantion of causation or transmission. For example, labor-saving technical change may result from and be embodied in a technically superior machine, such as a new model typewriter. Following the usual convention, the dot notation represents derivatives with respect to time. The two-factor case is used to introduce the concepts.

constant factor prices:

$$\frac{\partial(K/L)}{\partial t}\frac{1}{(K/L)} \left| \begin{array}{l} >0 \text{ labor-saving} \\ =0 \text{ neutral} \\ < \text{capital-saving} \\ \text{factor prices } \left(\frac{W}{R}\right) \text{ constant} \end{array} \right. \tag{6}$$

Salter [462] argues convincingly in favour of the above definitions at the micro level of the firm or single industry, but the distribution of income at the economy-wide level was the main object of interest in the earlier studies of Hicks [225] and Robinson [417]. Hicks [225, p. 121] originally argued that "we can classify inventions accordingly as their initial effects are to increase, leave unchanged or diminish the ratio of the marginal product of capital to that of labor. We may call these inventions 'labor-saving,' 'neutral' and 'capital-saving.'"

But to compare situations before and after a change in technique, something must be held constant. For aggregate analysis, factor endowments, rather than factor prices, may be regarded as fixed. This has led to widespread acceptance [204, 206, p. 213] of definitions similar to that of Kennedy and Thirwall [272, p. 20], who assert that "Hicks defined a 'neutral' invention as one which *with given factor proportions* raised the marginal product of labor in the same proportion as the marginal product of capital."

Economic interpretation is again simple and appealing. A labor-saving innovation makes labor in some sense more plentiful relative to capital than it was previously, with the result that the marginal product of labor must fall relative to that of capital. Since the equilibria shown in Figure 2.1 require that the ratio of marginal products be equal to the ratio of factor prices, this is equivalent to a rise in the price of capital relative to that of labor.

Bias and neutrality can again be formally defined in terms of the proportional change in the ratio of marginal products as a *constant factor ratio.*

$$\frac{\partial\left(\frac{F_k}{F_l}\right)}{\partial t}\frac{1}{\left(\frac{F_k}{F_l}\right)} \left| \begin{array}{l} > \text{labor-saving} \\ =0 \text{ neutral} \\ < \text{capital-saving} \\ \text{factor ratio } \left(\frac{K}{L}\right) \text{ constant} \end{array} \right. \tag{7}$$

Salter [462, pp. 32–33] argues that this definition reverses the reasoning of the first approach but results in the same division of innovations into labor- and capital-saving categories. Binswanger [53, p. 43] notes that Eq. (6) is actually equal to Eq. (7) multiplied by the elasticity of substitution. So, although the two definitions have varying economic implications, formally they differ only by a scalar multiple. Both of these statements of the relationship require careful qualification, for they are true only for a restricted class of production functions. Early critics of the "Hicks–Robinson" classification (Eq. (7) above) argued that it is only for linear homogenous production functions that marginal productivities depend only on factor ratios and are independent of the level of output [61]. Alternative definitions of Hicks neutrality and the relationship between them implied by different restrictions on the production function are investigated by Blackorby, Lovell and Thursby [60].[17]

A third definition of neutrality follows from those above. If, at constant factor prices, the ratio of capital to labor increases, the ratio of capital's share relative to that of labor, (rK/wL), must also increase. Alternatively, if at a constant factor ratio the price of capital increases relative to the price of labor, capital's relative share must increase. Thus by Eq. (6) or (7), technical change must be labor-saving if capital's share increases, neutral if shares remain constant, and capital-saving if capital's relative share falls.[18]

The attraction of the factor shares definition is that it generalises easily to handle many factors of production. A single measure of

---

[17] They pointout that Hicks did not hold factor ratios constant, but required that the firm should remain in a position of "internal equilibrium." This is interpreted to mean "expansion path preserving." If the product function is not homothetic, a shift from A in Figure 2.1, to a point like F on the non-linear expansion path OFA is Hicks-neutral by their definition but labor-saving according to the fixed factor proportions approach.

[18] If marginal products and factor prices are equal, the factor share definition can be written in terms of output elasticities. If Y is output, the output elasticity for capital is the ratio of the marginal to average product, or capital's share in output,

$$\left(\frac{\partial Y}{\partial K} \cdot \frac{K}{Y} = \frac{rK}{Y}\right).$$

Robinson's [417] diagrammatic analysis exploits the relationship between marginal and average product.

bias for each factor[19] is given by

$$\frac{\partial S_i}{\partial t}\frac{1}{S_i}\left|\begin{array}{l} >0 \text{ factor } i \text{ saving} \\ =0 \text{ neutral} \\ <0 \text{ factor } i \text{ using} \\ \text{relative factor prices constant} \end{array}\right. \tag{8}$$

where S_i is the share of factor i (in total costs) [53, p. 21]. Empirical applications of this measure are becoming increasingly common in multiple input studies and will be discussed in Section 2.3.

2.1.2. Harrod neutrality and the theory of economic growth. In the long run, neither factor prices nor factor input ratios can realistically be held constant. In growth models, the equivalent of the static equilibrium of the last section is the steady state, in which all variables grow at constant rates. Harrod's definition of neutrality is compatible with the existence of steady state growth,[20] allowing technical change to be incorporated in standard growth models without disturbing the balance between labor and capital. This is achieved by exploiting the fact that purely labor-saving (augmenting) technical change[21] is analogous to population growth. On the balanced growth path of the neoclassical model, all variables grow at the same rate as the exogenously determined growth rate of population. If this rate is $\dot{L}(t)/L(t) = n$, then exogenous labor-augmenting technical change at rate $\dot{B}(t)/B(t) = m$ can be incorporated by redefining labor in "efficiency units," $L = B(t)L$, that grow at rate $n + m$.

---

[19] Obviously if there are n factors, there will be n measured biases, whereas for the pairwise ratios of Eqs. (6) and (7) there will be

$$\frac{n!}{2(n-2)!}$$

combinations of bias parameters and associated difficulties of interpretation.

[20] It is in fact the only definition of neutral technical change that has this property. This was recognised by Robinson [417] and proved by Uzawa [548]. Jones [247, pp. 167–171] employs a simple heuristic argument to explain the problem.

[21] In terms of factor augmentation, Harrod neutrality is the Eq. (4) case, previously referred to as purely labor-saving under the Hicks classification. In Figure 2.1, if labor (L) is replaced by $\hat{L} = B(t)L$, so that labor is measured in efficiency units, what used to be Hicks neutrality becomes Harrod neutrality.

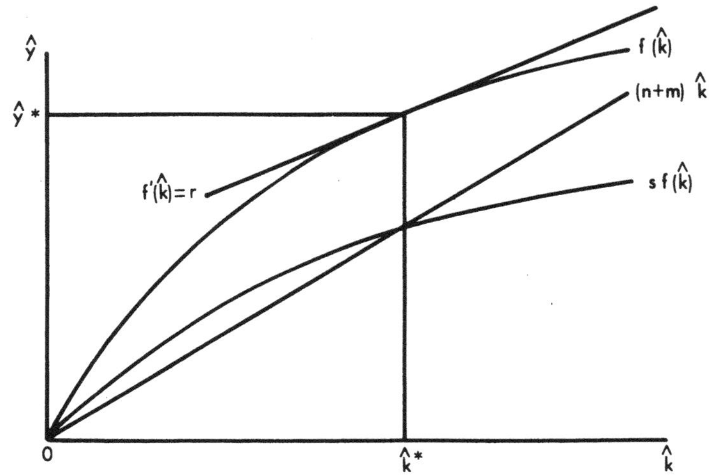

FIGURE 2.2 Technical change in the neoclassical growth model.

This proposition is shown in Figure 2.2, where the labor-augmenting specification of the production function

$$Y = F[K, B(t)L] \tag{9}$$

is assumed to exhibit constant returns to scale so that it may be written in the labor intensive form, $y = f(k)$ where $y = Y/L$ and $k = K/L$. Converting to efficiency units of labor, let, $\hat{y} = Y/B(t)L$ and $\hat{k} = K/B(t)L$. The standard neoclassical growth model diagram can then be defined in terms of efficiency units of labor, with a steady-state equilibrium at $\hat{k}^*$ where saving per effective worker, $(sf(\hat{k}))$, is equal to the growth of effective labor, $(n + m)$.

Modern adaptations of Harrod's definition of neutrality [247, p. 164] require that technical change should leave the marginal product of capital unchanged at a constant capital-output ratio.[22]

[22] For technical change to be both Hicks- and Harrod-neutral, the production function must be Cobb–Douglas. If the capital input and its price are unchanged relative to output, yet the input of labor is decreased, the rising wage must exactly compensate for factor shares to remain constant. The unitary elasticity of the Cobb–Douglas was originally intended to give this constancy of shares. To show Harrod neutrality in Figure 2.1, rotate the isocost line around point P until it is tangential to I_1 at point G. Then capital remains at $\bar{K}$, output is unchanged and so is the price of capital [204, p. 122].

The diagram shows this to be the case, since Y and K both grow at the same rate of $(n+m)$, and $\partial Y/\partial K = f'(k)$ is constant at the steady state equilibrium of $\hat{k}^*$. The terms $\hat{y}$ and $\hat{k}$ are also constant, but are defined in efficiency units. Hence, Y/L and K/L, in natural units, both grow at the rate of technical change, which is m. Thus, including a simple representation of technical progress in the model does not damage the harmonious results and gives conclusions that are closer to Kaldor's [258] "stylized facts" of economic growth.

Again, bias can be defined relative to neutrality for the Harrod classification of neutrality, but this is of little interest since there have been no growth-theoretic empirical investigations of biased technical change. Instead, Section 2.2.5 takes up the theoretical problem of explaining the systematic Harrod neutrality of technical change, since if this requirement cannot be justified, the conventional steady state approach to growth is hard to defend.

2.1.3. Extensions: two-sector growth models and other definitions of neutrality. Though the discussion above includes the case of several factors of production, further complications arise if more than one good or sector is considered. Then neutrality in aggregate depends not only on the direction of technical change in each sector, but also in the relative rates of change and the relative sizes of the sectors. As a result, technical change is affected by demand [149] and further complicated by changing relative prices.[23] Jones' [248] diagrammatic treatment is accessible to non-specialists and includes a summary of possibilities in the two-sector case plus references to earlier studies. Whereas Jones concluded that Hicks neutrality is "unlikely," Steedman's [513] model allows for inter-industry linkages and reaches the conclusion that under several plausible conditions, Hicks neutrality is in fact impossible. Chang [83] offers a detailed discussion of the relationships between Hicks, Harrod, and Solow neutrality[24] in the case of the two-sector model.

---

[23] This was recognized by Blaug [61], who considers that the effects of changing relative prices and commodity substitution constitute a "fatal objection" to the Hicks–Robinson classification of technical change.

[24] In terms of the factor-augmenting production function, this is the purely capital-augmenting case of Eq. (5).

2.1.4. Critique: technical change, factor substitution, and returns to scale. Hicks' definition of technical change was intended to distinguish it from factor substitution, another concept introduced in the same book [225] in which he also suggested that innovations may be "induced" by "a change in relative factor prices." The following section will show that much confusion over the induced innovation hypothesis is attributable to the more basic conceptual difficulty of separating technical change from factor substitution.

The neoclassical conventional wisdom defines the production function to be the boundary of the production set. It is thus the locus of *technically* efficient input combinations for differing levels of output, embodying all known techniques. The *economically* efficient point on the production function itself, or on the isoquant derived from it, depends on relative price ratios. Thus, factor substitution is a (costless and instantaneous) movement *along* an existing production function or isoquant, in response to a change in prices. By contrast, new technology, which *shifts* the isoquant or production function, is the product of research that requires time and consumes real resources.

This clear theoretical distinction may be a poor description of a more complex reality in which "*substitution* of real capital for labor in response to *innovations* which made the increased use of machinery and equipment technically feasible and economically profitable is even today the main instrument for actualizing productivity gains" [479]. But the basic problem is the isoquant itself. Why would a society with a high capital–labor ratio ever have available detailed knowledge of labor-intensive techniques of production, given that knowledge of production is costly? Rosenberg [432, p. 63] suggests that "the notion of a wide range of alternatives readily available, as implied by the drawing of smooth, continuous isoquants, is largely a fiction." If a firm has to commit resources to research and development to allow factor substitution, new knowledge is being created and the activity should be called technical change rather than factor substitution. Even if alternative techniques do exist, Rosenberg [432, p. 64] argues that "today's factor substitution possibilities, in other words, are the product of yesterday's technological exploration." David's [100, Ch. 1] "linear programming" approach similarly questions the validity of the distinction between factor substitution and technical change and stresses the importance of learning by doing and the localized nature of technical knowledge.[25] Atkinson and Stiglitz [18] argue that technical change would shift only that portion of the production function in the immediate vicinity of the factor ratio actually being used (above $\hat{k}^*$ in Figure 2.2). For example, would the firm "really want to raise productivity on handcarts as well as forklift trucks?" (p. 577).

To summarize, Nelson [353, pp. 64–66] notes that "if one drops the assumption that learning and doing are different activities, then the clear distinction between moving along a production surface and shifting the production function is smudged . . . . Rather than facing a sharply defined set of well-understood techniques with closely predictable inputs and outputs (and an abyss beyond), it would seem more plausible to characterize a firm as having a number of techniques that it can use with considerable confidence, others which might require a certain amount of research and development and learning-by-doing, and still other techniques about which the firm is even more uncertain and which likely would require even more resources and time before the firm could get them under effective control."

Apart from the conceptual problems considered above, serious difficulties arise in the empirical estimation of technical change. Indeed, a literature has developed on the impossibility of the simultaneous estimation of the elasticity of substitution and the biases of technical change [113, 114, p. 444; 468]. These papers also consider the problems of distinguishing technical change from returns to scale, an issue that has attracted attention since the exchange between Solow and Stigler [517]. Most contributions are considered in three survey articles [272, 349, 392], while recent developments are discussed by Dogramaci [121], Sato [468], and Sato and Suzawa [471].

Though recent contributions to the empirical literature covered in Section 2.3 show the progress that has been made since the early production function studies of technical change that imposed constant returns to scale, the problem persists, since a new

[25] David's critique of the methodology of neoclassical economics and that of Nelson and Winter appear in the final section of this paper as a prelude to the evolutionary models they propose as an alternative.

technology may allow economies of scale to be realized that were not previously attainable [392, p. 505]. Even the assumption of constant technology in cross section studies is invalid if diffusion is not instantaneous. Part 3 of this survey suggests that large units adopt new technology more quickly than smaller competitors. In such cases, cross section estimates of "increasing returns to scale" may actually be a measure of technical superiority.

The quality and meaning of estimates of biased technical change will also depend on the approach taken to the interdependent problems of aggregation, quality adjustment, and index numbers. These issues are covered in the surveys of technical change [272, 349, 392] but recent progress has been rapid [268]. Particularly, the realization that index number formulae can be derived explicitly from particular production functions has provided a powerful new basis for selecting index procedures [81]. A production function with suitable properties can be selected and the corresponding "exact" index derived. Diewert [117] argues in favor of flexible functional forms (that can provide a second order approximation of an arbitrary production function), calling index numbers that are exact for such functions "superlative."

These empirical issues are deferred while we move from the classification and measurement of exogenous technical change to the induced innovation hypothesis, which attempts to increase the explanatory power of economics by endogenising technical change. Binswanger [53, p. 13] provides a definition: "Models of induced innovation and empirical tests of such models are an attempt to discover the roles played by factor prices, goods prices, and other economic variables in determining the *rate* and *direction* of technical change."

2.2. Endogenous technical change: the induced innovation hypothesis

Economic explanations of the rate and bias of technical progress have been prominent in the study of economic history, at least since Mantoux's [320] classic study of the industrial revolution in Britain, which stressed "economic needs and the spontaneous efforts they call forth." A good sample of historical examples is used in Rosenberg [429] to explain how imbalances, bottlenecks, and expensive or troublesome labor stimulated particular mechanical inventions. Recent work on the industrial revolution in England [341] is more critical, pointing out that bottlenecks were frequently overcome by reallocation of factors, rather than invention.

2.2.1. Induced innovation in economic history. The historians' main contribution to induced innovation is to be found in the lengthy debate on British and American technology in the nineteenth century. Rothbarth's [439] original statement of "the labor scarcity hypothesis" attributed high labor productivity in the United States relative to Britain in the nineteenth century to the greater use of labor-saving equipment, caused by the relatively higher industrial wage rate in the United States. The high industrial wage rate in the United States was explained by the need to compete with high returns to labor in the agricultural sector, which resulted from the abundance of a third factor, land. Rothbarth's work and the more extensive study by Habakkuk [203] highlight many of the difficulties involved in the study of biased technical change.

Temin [538] points out that neither author offers a clear distinction between factor substitution and innovation; yet there is a considerable difference between more machinery per unit of labor and better machinery. In addition, Saul [474, p. 18] argues that "a careful distinction must be drawn between invention, innovations and diffusion." Best practice techniques could have been the same in the two countries but the United States have a better capital stock in aggregate due to a more rapid diffusion of innovations. Indeed, if the U.S. growth rate were greater and/or the durability of machinery lower, then embodied technical change would lead to a superior capital stock (but not automatically to a labor-saving bias) in the United States, regardless of the inducement mechanism driven by relative factor prices (see Williamson [567] for a discussion of this view). Ames and Rosenburg [11] include land, in the form of natural resources, in the manufacturing production function[26] and suggest that the "technological superiority" of U.S. industry was in fact dependent on "natural resource" intensive techniques. Christensen [90] follows this lead, arguing that labor-

[26] In Temin's [538] application of the three-factor, two-sector model, land is specific to agriculture and capital to manufacturing, with labor common to both.

saving, capital-intensive American technologies were resource-using, especially in the sense of exploiting the plentiful cheap horsepower that was available.

Though too brief to do justice to the literature on British and American technology in the nineteenth century, this summary serves to show that a rigorous theoretical framework is required to distinguish between factor substitution and the rate and bias of technical change. Empirical tests of the propositions raised in the debate are covered in the empirical Section (2.3) of this survey. For a comprehensive review of the debate see David [100, Ch. 1].

2.2.2. Microeconomic approaches. Confusion over the meaning of induced innovation can be traced directly back to Hicks' original, widely quoted statement: "The real reason for the predominance of labor-saving inventions is surely that which was hinted at in our discussion of *substitution.* A change in the relative prices of the factors of production is itself a spur to *innovation,* and to inventions of a particular kind—directed at economizing the use of a factor which has become relatively expensive" [225, pp. 124–125].

In later contributions, Hicks [228, Chs. 1 and 2] explains the mechanism whereby autonomous inventions provide the initial *impulse,* which would peter out due to labor scarcity raising wages but for the "children" of the original innovation, the labor-saving secondary innovations, *induced* by the original improvement. "But whether such 'induced inventions' were to be regarded as shifts in the Production Function, or as substitutions within an unchanged Production Function, was left rather obscure" [228, Ch. 1, p. 2].

As a result, the induced innovation hypothesis was not readily accepted by economists. In his survey of process innovations, Blaug [61] refers to "the troublesome notion of innovations induced by changes in factor prices—this would seem to involve factor substitution, not technical change." Indeed, Salter's [461] refutation of inducement was favorably received by economists [156, p. 337; 429]. He argued that, "at competitive equilibrium, each factor is being paid its marginal value product; therefore all factors are equally expensive to firms" [461, p. 16]. Factor substitution ensures that this efficiency condition will be re-established so that no factor is ever "relatively expensive." Thus, "the entrepreneur is interested in reducing costs in total, not particular costs such as labor costs or capital costs. When labor costs rise, any advance that reduces total cost is welcome, and whether this is achieved by saving labor or capital is irrelevant" [461, pp. 43–44].

Clearly, Salter does not reject factor substitution and it is the problem of differentiating this from induced innovation that underlies his objection to the concept. Intending to keep logically separate the technological possibilities and the economic forces that determine the techniques actually in use [462, p. 15], "Salter defined the production function to embrace all possible designs conceivable by existing scientific knowledge and called the choice among these designs 'factor substitution' instead of 'technical change'" [222, p. 86]. As Rosenberg [433, p. 65] points out, factor substitution then swallows up much of technical change since the production function is no longer a set of blueprints on the shelf, but is also the "much wider range of techniques which could be designed with the current stock of knowledge" [461].

Salter's rejection of induced innovation is semantic, showing that isoquants can be defined so as to leave no room for technical change or none for factor substitution. The second extreme is exemplified by Brozen [74, p. 88], who considered the distinction between known but previously unused techniques and new technology to be so problematic operationally that he defined technological change as "any change in production methods in an enterprise or industry." Fellner, in a series of studies [148, 149, 150, 151, and 152], both expressed a view similar to that of Salter and rehabilitated the induced innovation hypothesis. He argued firstly, that when change occurs, even a perfectly competitive firm will find itself in a quasi-monopsonistic situation, facing a less than perfectly elastic supply curve for a factor that is in relatively short supply at the economy-wide level and will learn to direct innovative activity toward saving that factor. Secondly, even a perfect competitor will adapt to a persistent and discernible relative factor shortage. The *expectation* of future factor shortage is sufficient to generate a bias in inventive activity.

Ahmad's [2] rehabilitation of the inducement hypothesis does not require imperfect markets or expectations. The novelty of Ahmad's model is the introduction of the innovation possibility curve (IPC), which he defines to be the "envelope of all the alternative isoquants (representing a given output on various production functions) which the businessman expects to develop with the use of the available amount of innovating skill and time (assumed constant throughout this analysis)" [2, p. 347].

If relative factor prices change, factor substitution will occur in the short run, but over a longer time period there will be substitution along the IPC so that a new isoquant is created and selected.[27] However, Admad assumes that over this same longer period, research and development expenditures will have neutrally[28] shifted the IPC closer to the origin, so that the new equilibrium will be on an isoquant associated with a new IPC($t + 1$).[29] If the IPCs have the properties of input homothetic isoquants (see footnote 17), then changing actual relative prices will induce a factor-saving bias. This need not be the case, Ahmad cautions; the IPCs could be drawn to show an innate factor-saving bias in innovation possibilities.

Ahmad's model has been extended to encompass several inputs and applied to the problem of agricultural development by Hayami and Ruttan [222], who propose that "technology can be so developed as to facilitate the substitution of relatively abundant (hence cheap) factors for relatively scarce (hence expensive) factors in the economy" (p. 73). Their model is developed by utilizing the identity

$$Q/L \equiv (A/L)(Q/A) \tag{10}$$

where Q is output, L is labor, and A is land.

Land area per worker (A/L) can be increased by technical improvements in machinery and equipment, which allow power to be substituted for labor. This process may be called mechanical technical change. Similarly, biological advances, such as high-yielding, fertilizer-responsive seed varieties, raise the average prod-

[27] Induced innovation in Ahmad's model becomes factor substitution under Salter's definition of the isoquant.

[28] Salter's rejection of induced bias is equivalent to Ahmad's assumption that the IPCs shift Hicks neutrality [204, p. 126]. Both are unwarranted assumptions. In particular, Ahmad postulates neutrality of the shift of the IPC also in the sense that it is independent of the technique actually used in the previous period [368, p. 212].

[29] To avoid repetition, Ahmad's diagram is omitted; see the diagrammatic explanation below of the Hayami and Ruttan model, which follows Ahmad's reasoning.

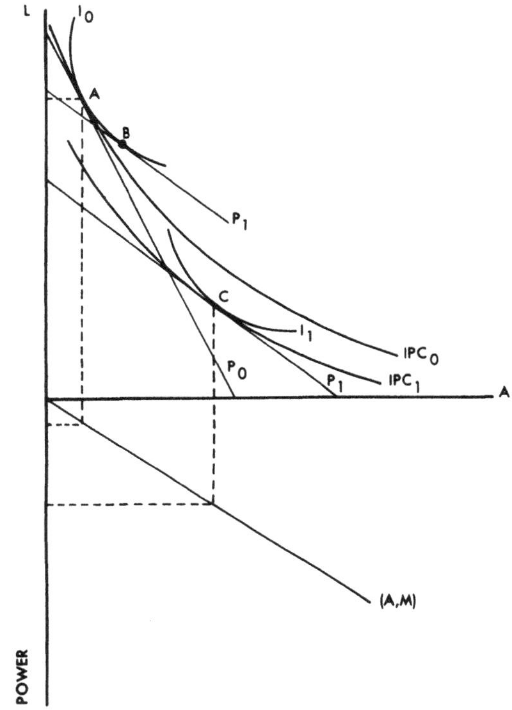

FIGURE 2.3a Factor prices and induced mechanical technical change.

uct of land (Q/A) and may be referred to as biological/chemical technical change.[30]

In Figure 2.3a the initial price ratio P_0 is tangential to the IPC, IPC_0 at point A, and has led to the development of a particular technology (the reaper, for example) described by the isoquant I_0. (At another price ratio, some other isoquant, along IPC_0, would have been developed.) When the factor price ratio changes to P_1, factor substitution allows land to replace labor until a new tangency is reached at point B. However, over a period of time sufficient to allow for the development of new techniques, inventions suited to the new factor price ratio will appear (the combine harvester,

[30] Griliches [194, pp. 241–245] has suggested that these two processes may be regarded as "somewhat independent (of each other), at least over a certain range."

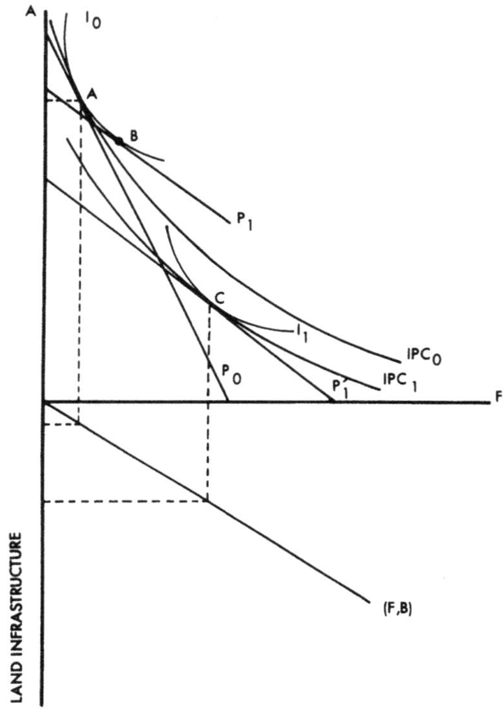

FIGURE 2.3b Factor prices and induced biological/chemical technical change.

perhaps), represented here by the isoquant I_1. The new production point, C, on I_1 lies on a new IPC, labeled IPC_1, which is closer to the origin than the original IPC, the extent of the shift being a function of the level of the research and development budget.[31]

The new technology, I_1, allows a higher land-labor ratio but does require a greater input of power per worker, as is shown by the line (A, M), which implies complementarity between land and power.[32]

Similarly, biological/chemical technical change is shown in Figure 2.3b with the initial equilibrium at point A, where the isoquant I_0 and the IPC, IPC_0 are tangential to the price ratio P_0. The invention of fertilizer-responsive high-yielding varieties, in response to a fall

[31] Though the model explains the direction of the factor-saving bias of technical change, the rate of technical progress remains exogenous.

[32] The diagram is clearly a simplification and is not intended to represent fixed factor proportions between land and power.

in the relative price of fertilizer, to P_1, is represented by the isoquant I_1. This isoquant is one of a family of such curves, for which IPC_1 is the envelope curve. At the new equilibrium point C, fertilizer input per acre is increased but high-yielding varieties require better water control and land management. This complementary relationship between land infrastructure and fertilizer is implied by the line $[F, B]$.[33]

To simplify the presentation, Figure 2.3a treats the impact of advances in mechanical and biological technology on factor ratios as if they are independent, but biological technical change in Figure 2.3b reduces the land input per unit of output and will thus change the land-labor ratio in Figure 2.3a. Thirtle [540] shows that the assumption that the production function is separable is sufficient to allow a diagrammatic representation that takes account of these interactions and thus avoids specification errors in the empirical tests discussed later. Kaneda [264] has argued in favour of separability (denoted by :) between labor (L) and machinery (M) and land (A) and fertilizer (F) as in Eq. (11).

$$Y = f[(L, M, Tm):(A, F, Tb)], \tag{11}$$

where Y is output, Tm represents mechanical technical change, and Tb represents biological technical change.[34]

Equation (11) leads directly to a simplification[35] of the Hayami

[33] This account is close to that in Hayami and Ruttan [222]. The original version in their first edition is more simplistic. That of Ruttan, Binswanger, Hayami, Wade, and Weber [455] adds further complexity by differentiating between the short run (in which existing capital leads to nearly fixed proportions); the long run (in which substitution along the neoclassical isoquant occurs); and the secular period (in which, given the current state of scientific knowledge, a set of IPCs can be developed, each corresponding to a different research budget).

[34] Definitionally, the function is separable if,

$$\frac{\partial\left[\frac{F_i}{F_j}\right]}{\partial[X_k]} = 0 \quad \text{for all} \quad i, j \subset N \quad \text{and} \quad k \not\subset N.$$

where, for this two-group case, N denotes either group of factors, and F_i, F_j are the marginal products of X_i and X_j. Hence, separability requires that the rate of substitution between two factors in one group should be independent of the level of factor inputs in the other group.

[35] The model presented here is a simplification in the sense that it requires only conventional neoclassical isoquants that shift Hicks neutrally. The same four-quadrant approach can obviously incorporate IPCs and non-neutral technical changes in the spirit of the Hayami and Ruttan model [541].

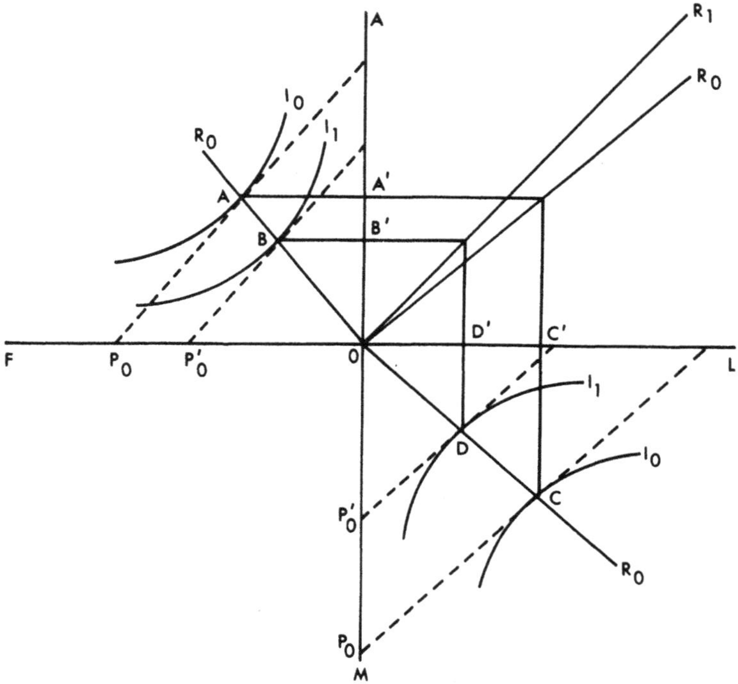

FIGURE 2.4 A neoclassical reformulation of the Hayami and Ruttan model.

and Ruttan model that can be represented by a single diagram. Figure 2.4 exploits the separability assumption, showing labor, machinery, and mechanical *TC* in the southeast quadrant, while land, fertiliser, and biological/chemical *TC* are represented in the northwest quadrant. The relationship between land and fertiliser shows the initial equilibrium at point *A*, where the unit isoquant I_0 is tangential to the relative price ratio P_0. Biological/chemical *TC* is represented by a neutral shift of the isoquant to I_1, which at constant factor prices results in a new equilibrium at point *B*. The proportional reduction in the input of land is measured by $A'B'/OA'$.

Similarly, the initial equilibrium in the labor–machinery quadrant is at point *C*, where the unit isoquant I_0 is tangential to the factor-price ratio P_0. Mechanical *TC* is represented by the neutral shift of the isoquant to I_1, resulting in a new equilibrium at point *D*.

The proportional reduction in the input of labor is measured by $C'D'/OC'$.

Figure 2.4 represents a situation in which mechanical *TC* reduces the input of labour more than biological TC reduces the input of land, ($C'D'/OC' > A'B'/OA'$), with the result that the land/labor ratio rises from R_0 to R_1 (in the northeast quadrant). Even though the mechanical and biological technical changes are constrained to be Hicks-neutral (in order that they may be measured by single parameters), their effect on the land/labor ratio will be non-neutral unless the two changes are equal. Thus, allowing for the interaction of the two types of technical change permits changes in the land labor ratio even at constant factor prices. This possibility does not exist in the Hayami–Ruttan model and is not allowed for in their tests, discussed in Section 2.3.

Though the effects of research enter the Ahmad and Hayami–Ruttan models by way of the shifting of the IPC toward the origin, neither approach makes an explicit attempt to model the research process. For the case of one output and two inputs, in the context of the Cobb–Douglas and CES production functions, Kamien and Schwartz [260] include a "research production function," $I(a_1', a_2') = M$, where a_1' and a_2' are the time derivative of the function parameters and M is the size of the research budget (assumed constant). Their model, which maximizes the present value of the future stream of net profits, shows that the optimal direction of disembodied technical change depends on the initial technology, relative factor prices, and the relative costs of acquiring different types of technical change. In a later paper, Kamien and Schwartz [262] allow for a variable research budget and decreasing or constant returns to research expenditure. Subject to the assumptions of the study, the rates of neutral and non-neutral technical change vary inversely with the costs of these changes, directly with the responsiveness of the cost function to that type of change, and directly with firm size. They also find a long-run tendency toward Hicks neutrality and failure of a myopic policy of instantaneous maximisation (as opposed to a dynamic solution) to achieve either the optimal rate or bias of technical change.

The contributions of Binswanger [50, 54, 55] adapt Evenson and Kislev's [132, 133] stochastic model of applied technological research to the problem of induced innovation. Using seed technology as an example, research is viewed as a sampling process from a distribution of potential yield increases that depend upon nature, the state of basic science, and plant breeding techniques. Ex ante, the expected payoff is $E(\Delta y_m) = h(m, \mu, \sigma)$ where y_m is the largest yield increase, m is the sample size, and μ and σ are respectively the mean and variance of the distribution.[36] Binswanger [50] follows Kamien and Schwartz [261] in using a factor-augmenting form of the production function, with the change in the augmentation parameters a function of research effort. Then, for the case of two inputs (hence two augmentation coefficients A and B) and two research processes m and n, innovation possibilities may be specified as

$$A^* = M(m)\alpha^m + M(n)\alpha^n \quad (12)$$

and

$$B^* = M(m)\beta^m + M(n)\beta^n, \quad (13)$$

where A^* and B^* are proportional changes in A and B, $M(i)$ are scale functions, and α^i and β^i are the productivity coefficients of research to reduce A and B, respectively.[37] The functions are used to maximize the discounted present value of the profit function, subject to a variable research budget. Though Binswanger considers many cases, the main results derived from a maximization model of discounted expected costs and benefits may be summarized as follows:

1) Any rise in the expected present value of the total cost of a factor will lead to an increased allocation of resources to the research activity that most saves that factor.

2) A rise in the cost of research that saves a particular factor or a decline in the productivity of that research will reduce the allocation to that line of research, and hence bias technical change in the direction of the other factor.

---

[36] If the parameters are fixed, payoffs to applied research will be rapidly exhausted. Basic research is required to change the state of scientific knowledge and hence μ and/or σ so that the process may be continuous.

[37] Note that each research process reduced both augmentation coefficients, but in different proportions. Also, these functions are not specific to the factor-augmenting form of the production function. Indeed, Binswanger [55] retains these same functions, but with A and B as parameters of the production function.

3) With no budget constraint on research activities, a rise in the value of output (due to greater output or higher price) will increase the research budget and hence the rate of productivity growth.

Ex post, none of these results are particularly surprizing, but they do clarify several controversial issues in the earlier literature. The first result shows that it is neither factor prices alone, as in the Ahmad and Hayami–Ruttan models, nor the expectation of rising wage rates, as in Fellner's study, nor factor shares, as in the Kennedy–Weizsacker–Samuelson models (see Section 2.2.5), but the present value of the (expected) factor costs that determines the bias of the research mix and hence of technical change. Moreover, Binswanger shows that the IPC should not be viewed as the "scientific frontier," as no firm will intentionally drive the returns to research to zero. Profit maximization requires that the effort cease when the marginal cost of research is equal to the marginal product. The third result listed above implies that for both society and the firm, more research resources should be concentrated on commodities with higher prices and larger markets. Thus with respect to the rate of technical change, the model provides theoretical support for the importance of demand in inducing technical change (see Part 1 on the rate of technical change).[38]

In his criticism of the Kennedy and Ahmad approaches to induced innovation, Nordhaus [368] employs a family of "isotechs." The neoclassical isoquant is the zero isotech, and the set of all techniques attainable at a given cost C is the C isotech, which is analogous to Ahmad's IPC except that the rate of technical change is endogenized. There is a set of isotechs that shift closer to the origin as research costs, C_i, increase. Thus both factors can be saved with a larger research budget. The model is not ahistorical, since the actual technique employed on the isotech in the first period will determine the shape of the isotech in the next period. McCain [326] extends this approach to investigate the scale and durability of "new designs" in addition to the capital and labor intensities. Wyatt [576] also extends the isotech approach, showing

[38] Binswanger [54, 55] studies many other issues, including the effects of market structure, induced bias in the output mix, the effect on the bias when the technical change is embodied in an intermediate input, demand conditions, a budget constraint on total research resources, and the effect of varying degrees of patentability for different factor improvements.

that when new technology is embodied in capital equipment, the rate of technical change will itself affect the factor-saving bias.

2.2.3. Criticisms of the microeconomic approach to induced innovation. Some theoretical limitations of earlier models of induced innovation, discussed in the previous section, are resolved by later contributors, particularly Binswater, but other deficiencies remain unanswered. Hacche [204] argues that the treatment of uncertainty and expectations is inadequate, but his most serious objection is to the Hicks-neutral shifting of the IPCs specified by Ahmad. A similar weakness underlies the growth-theoretic models of technical change, discussed below.

Elster [127, pp. 102–103] attributes the appeal of the Hicksian argument to "an easily committed logical fallacy." If wages are rising relative to capital costs, then for entrepreneurs collectively, labor-saving innovation seems to be appropriate. But labor-saving innovation will reduce wage rates (due to factor substitution), and since entrepreneurs act individually, not collectively, "the proposed explanation fails." This point may amuse logicians but has little to do with economic behaviour in a world where wage-rental ratios have continued to rise over time. Binswanger [54, p. 91] argues specifically that for both society and the individual firm it makes sense to take factor prices into account in determining the amount and direction of research effort.

Most criticism of induced innovation has centered on the practical validity of the market-price-based Hayami–Ruttan model as a foundation on which to build a theory of agricultural development. Rosenberg [429] suggests instead a theory of induced innovation based on the "obvious and compelling need" to overcome the constraints on growth of production or of factor supplies. (See also Davidson's [104] empirical study of this proposition). However, Hayami and Ruttan [221] argue that technical imbalances or bottlenecks should be reflected in an operationally meaningful fashion in terms of relative factor scarcities signalled by market prices. Indeed, Timmer (see footnote 18 in Hayami and Ruttan [220]) suggests that the constraints that give rise to the "obvious and compelling need" are, in a linear programming context, the dual of the factor prices in the Hayami–Ruttan model (provided market failure is ruled out).

Unfortunately, there is a low probability that the conditions required for efficient competitive equilibrium will be fulfilled in developing countries, as was pointed out by Beckford [31]. Especially when risk is taken into account, low-income agricultural producers may not be profit maximizers and any divergence between private and social cost may distort the rate and direction of technical change.[39]

In an early contribution to what has become known as the "structuralist theory" [45, p. 209], de Janvry [109] incorporates price distortions in a model similar to that of Ahmad [2] and shows that socially optimal innovations may not be developed in such cases. This approach has been criticized by Mueller [347], who argues that removing the factor price distortion is no solution, since the IPC does not exist in developing countries without effective agricultural research institutions. Instead, there is only a labor-intensive traditional isoquant and a technically superior, capital-intensive modern isoquant "transplanted from other countries." Removal of the price distortion results in a minor substitution of labor for capital in the modern sector, but the dual economy persists.[40]

Moreover, inequalities in land ownership and farm size are likely to generate dual technologies. Griffin [188, p. xiii] argues that in LDCs there will not be one set of prices, but that different groups may face radically different sets of relative factor prices depending upon their economic position and political power. Adding duality to the induced innovation model, Grabowski [184] shows that large landowners with access to credit may have an incentive to pursue technological developments that require non-labor inputs, such as fertilizer and chemicals, which are less available to small farmers.

---

[39] Though this is clearly a serious issue, justifying the following discussion, in some sense the real problem is the distortions themselves, not the inadequacy of the theory of induced innovation. Distortions clearly have biased technical change, for example, in the case of rice in Japan during the 1960s and 1970s and tobacco in the United States [452, pp. 88–90, 338–340]. Public research institutions could tailor research to efficient shadow prices, but if farmers face distorted relative prices, the research output will not be appropriate to farmers' perceived needs [451, p. 24].

[40] The concept of the "dual economy" has taken on several meanings. Here it can be taken to mean the coexistence of modern and traditional techniques and the lack of unique factor prices. Mueller implies that dual technologies may be the inevitable outcome of the technology gap and the international diffusion of technology rather than the result of duality in factor prices.

Thus, unequal access to the inputs necessary for successful implementation of green revolution technologies can cause a worsening of the distribution of income even if the technologies themselves are inherently scale-neutral.[41]

De Janvry [110] emphasizes that the same inequalities of economic, social, and political power will distort the research activities of public sector institutions in favour of the dominant farm interest.[42] When the inputs of public research institutions are viewed as public goods, demands for particular lines of research depend on the expected payoffs to conflicting interest groups. The supply of innovations will depend on the political and bureaucratic structure, while socioeconomic position determines the actual payoffs. Guttman [201] further extends the public goods approach in a model in which agricultural research funds are allocated according to the votes of interest groups. The model explains the allocation of United States agricultural research funds for 1969. In an empirical analysis of the provision of extension services to Indian villages, Guttman [202] finds "political variables" to be important in addition to efficiency criteria.

A second, separate line of criticism questions the ability of such a broadly based theory to provide genuine research policy guidelines to suit the diverse situations of developing countries. Biggs [43, p. 22] argues that neither the induced innovation theory nor the structuralist theory "analyses the actual decision-making and behavioral processes within research institutions that generate and promote new technologies." The institutional approach of Biggs [45] stresses the importance of imperfect bureaucratic structures, institutional environment, communications, linkages, feedback mechanisms, dependency, and control in agricultural research systems, both at the level of the formal research institutions and at the level of non-formal on-farm research and development.

---

41 This possibility has attracted considerable attention. Hazell and Anderson [223] provide an up-to-date review of many contributions and offer explanations for the divergence of conclusions.

42 The development of irrigated rice varieties by IRRI might be taken as evidence of such a bias, since farmers with irrigated land are assumed to be more prosperous than those who rely on rainfall (see Biggs [43], p. 27).

2.2.4. Induced institutional change. Though Hayami and Ruttan [219] focus on induced technical change in agriculture, they also consider institutional innovations, because much technical change had been produced by public sector institutions [450, p. 32]. The importance of institutional change was stressed by Polanyi [398], who maintained that institutional rather than technical change is the dynamic source of economic development. Following this line of thought, North and Thomas [370, 371] attribute the major sources of Western economic growth to changes in the institutions whose rules govern property rights, with the changes being brought about by the pressure of population against increasingly scarce resource endowments. Focusing on more recent economic history, Schultz [484] identifies the "rising economic value of man" during the process of economic development as the primary cause of institutional change.

Defining institutional innovation broadly so as to encompass organizational change (property rights and markets, as well as agricultural research and extension), Ruttan [449] follows de Janvry [110, 111] in stressing the interdependence and interaction of technical and institutional changes.[43] Institutional change may be induced by the *demand* for more effective institutional performance required for economic development, or it may result from advances in the *supply* of knowledge about social and economic behaviour, organization, and change.

Furthermore, Ruttan [449] has argued that sources of demand and supply for technical and institutional change are essentially similar. Hayami and Ruttan [222] identify two major sources of change in the demand for institutional change: firstly, the response to disequilibria in the allocation of the new income streams resulting from technical change, and secondly, the impact of changes in

[43] De Janvry and Dethier's [112] "structuralist theory" of induced innovations stresses the importance of institutional forces in modifying the market outcome. Ruttan [449] summarizes the findings of Ruttan and Binswanger [453] on the green revolution by observing that the wide diffusion of an institutional innovation—the socialization of agricultural research—has generated technical change that in turn has created new income streams and disequilibria that are a powerful source of further institutional change. As a result, property rights in land, as well as many other institutional arrangements, are being modified [457].

resource endowments and relative factor prices.[44] On the supply side there are also two major sources of change: firstly, the organization of group action to supply public goods, and secondly, advances in knowledge in the social sciences and related professions, which reduce the cost of institutional innovation.[45]

Though Hayami and Ruttan [222] have extended the induced institutional change model to include cultural endowments and Ruttan [453] has added a case study of the direct payment approach to United States farm income support, the model remains incomplete and difficult to test empirically.[46]

2.2.5. Growth theoretic approaches. The apparent importance of technical change in explaining economic growth (see Section 2.1) led to attempts to incorporate endogenous technical progress in modern growth models. The innovation possibility frontier (IPF) introduced by Kennedy [269][47] resulted partly from the author's dissatisfaction with the neoclassical production function, as did Kaldor's [257] earlier "technical progress function."[48] The IPF offers a theory of induced innovation and distribution (one aim was to explain the constancy of factor shares) that can be viewed as independent of the neoclassical production function.[49] It does not

[44] Hayami and Kikuchi's [217] Philippine study provides an example of both demand side sources. They argue that rapid population growth, combined with irrigation and high-yielding rice varieties, has led to institutional innovations in the form of subleasing and labor contracts that require weeding services as a precondition of participation in the harvesting operation. See also Feeny [144] and Feeny's [145] response on delayed irrigation projects in Thailand, in which he suggests the importance of politics and "returns to the men in government."

[45] Hayami and Ruttan [222] and Ruttan and Hayami [456] cite the communal arrangements in Japanese villages, designed to prevent depletion of common property, as an example of the first case and the role of social science knowledge in the design of more efficient commodity markets, land tenure institutions, credit, and marketing arrangements as examples of the second.

[46] However, an attempt to test the hypothesis suggested by Schultz [484] has been made by Stauffer and Blase [512].

[47] A similar model was simultaneously developed by von Weizsacker, who is accorded joint credit in much of the literature.

[48] Kaldor's approach is not pursued here, as it is the rate of technical change that he endogenized in a way that avoids difficulties by drawing no sharp distinction between techncial change and capital accumulation, on which its rate depends. Arrow's [15] "learning by doing" is similar in depending on investment and concentrating on the rate of change.

[49] The production function is made explicit in the explanation offered here, which is attributable largely to Jones [251, Ch. 8].

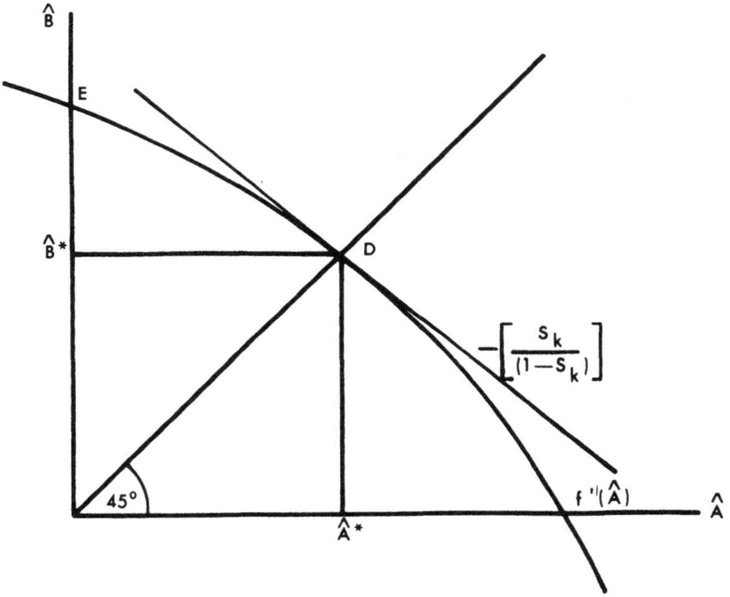

FIGURE 2.5 The innovation possibility frontier.

rely on changing relative factor prices, thus avoiding confusion between factor substitution and technical change.

In a two-sector model, technical change is assumed to occur only in the consumer goods sector, the rate of interest is constant, labor is homogenous, there is perfect competition, and the production function exhibits linear homogeneity. Using the factor-augmenting representation of Section 2.1, the proportional rate of reduction of the labor input due to technical change, $\hat{B} = \dot{B}(t)/B(t)$, can be raised only at the expense of less capital augmentation, $\hat{A} = \dot{A}(t)/A(t)$. Higher rates of labor augmentation require increasing sacrifices in capital augmentation. Thus the technology frontier (IPC) has the properties, $\hat{B} = f(\hat{A})$, where $d\hat{B}/d\hat{A} < 0$ and $d^2\hat{B}/d\hat{A}^2 < 0$, as shown in Figure 2.5. The entrepreneurs' objective is to maximize the one-period reduction in unit costs, $(\hat{C})$, which depends on the technical coefficients weighted by the factors' shares in total costs $(S_k, S_l = (1 - S_k))$. The objective function is

$$\hat{C} = S_k\hat{B} + (1 - S_k)\hat{A}, \tag{14}$$

and optimality is attained where

$$d\hat{B}/d\hat{A} = f'(\hat{A}) = -(S_k/(1-S_k)), \tag{15}$$

which is the point of tangency between the IPF and the factor share ratio at point *D* in Figure 2.5.

Figure 2.5 shows that a relatively high share of capital in total cost will lead to a greater value of capital augmentation, $\hat{A}^*$ (where * indicates the cost-minimizing solution), than will a lower relative share. However, if technical change is not Hicks-neutral, ($\hat{A}^* \lesseqgtr \hat{B}^*$), then the weights, S_k and $(1-S_k)$, will change in the next period so that the economy converges asymtotically to an equilibrium where technical change *is* Hicks-neutral ($\hat{A}^* = \hat{B}^*$). Thus factor shares remain constant at the levels determined by the slope of the IPF on the 45° line in Figure 2.5. The slope of the IPF at this point indicates "the fundamental technological bias in innovation possibilities."

If technical change occurs in the investment goods sector as well,[50] equality of $\hat{A}^*$ and $\hat{B}^*$ will not result in constant factor shares. The share of capital in total costs (S_k) will fall continuously as technical change in the investment goods sector lowers the price of capital supplied to the consumption goods sector. In this case Kennedy [269] shows that assuming a constant rate of interest and an elasticity of substitution of less than unity, there will exist a unique, globally stable balanced growth equilibrium, characterized by Harrod-neutral or labor-augmenting technical change and constant factor shares.[51] Intuitively, if *r* is fixed and the relative shares

[50] This gives a two-sector, one-good model, "or what is the same, to restrict the analysis to a one sector model" [156].

[51] This statement synthesizes several contributions. Kennedy [269] assumes a constant rate of interest and derives the Harrod neutrality result. Samuelson's [463] assumptions differ and he does not reach the Harrod neutrality result. His analysis introduces a factor-augmenting production function and facilitates the investigation of the stability conditions, especially the crucial importance of $\sigma < 1$. Any tendency for the share of labor to rise will make labor-augmenting technical change more profitable, and the introduction of technology with a labor-saving bias will reduce labor's share provided that the elasticity of substitution is below unity. Further analysis of the stability conditions can be found in Drandakis and Phelps [122], Wan [557], and Chang [84, 85]. Using the notation of Section 2.1, Drandakis and Phelps define the bias (*B*) in terms of the elasticity of substitution and the augmentation parameters as

$$B = [(1-\sigma)/\sigma]\left[\frac{\dot{B}(t)}{B(t)} - \frac{\dot{A}(t)}{A(t)}\right].$$

are written as

$$\frac{S_k}{S_l} = \frac{rK}{1 - rK}, \tag{16}$$

capital's share will fall relative to that of labor until $\hat{A} = 0$ and capital augmentation ceases at point E in Figure 2.5. Drandakis and Phelps [123] and Samuelson [464] integrated Kennedy's IPF into the standard one good neoclassical model, giving the required outcome. As Wan [557, p. 223] and Jones [247, p. 200] observe, the Kennedy approach offers an escape from the necessity of *assuming* Harrod neutrality for balanced growth to be possible; rather, it is a result of the model.

Clearly, maximizing the instantaneous rate of unit cost reduction may be shortsighted, but the Samuelson [463] version of the model minimizes unit costs T periods from the present and von Weizsacker [555] minimizes the total discounted cost of the future output stream. The more serious short-coming, that technical change is costless or results from "exogenously supplied inventions" [122, p. 11], is tackled by von Weizsacker [555] by allowing the firm to allocate a variable amount of "indirect" labor to research and development.

When investment in research and development is incorporated in the model, the rate of technical change may be determined, along with its direction. The optimal rate of technical change is determined by Uzawa [548] and Phelps [393, p. 139], who state the "Golden Rule of Research" as calling for "equating the (marginal) rate of return from research to the growth rate." Nordhaus [366, 367] incorporates Uzawa's result in an induced innovation model, so that the IPC is pushed outward in a homogeneous fashion, with the magnitude of the shift a function of the level of research and development investment. Nordhaus [367, pp. 107–108] arrives at the same optimality condition as Uzawa, which he compares to the conclusion of Phelps and von Weizsacker before stating his own golden rule of technological change.

In an alternative model that includes both the rate and bias of technical change, Conlisk [93] allows fractions of both employed capital and employed labor to be allocated to the capital and labor-augmenting research sector (the allocation varying only with the capital/labor ratio to preserve linear homogeneity). Two main conclusions follow. Firstly, in contrast to the simple neoclassical growth model (with exogenous growth of labor) where the equilibrium rate of growth is not affected by the savings rate, including labor as an "endogenously produced" factor does make the equilibrium growth rate a function of the savings rate. Secondly, in contrast to the models of Samuelson [464] and Drandakis and Phelps [123], which have a fixed rate of technical change, the bias in technical change need not be Harrod-neutral.

Indeed, Conlisk suggests that this odd feature of the neoclassical model may vanish as technical change is made increasingly endogenous.[52] Hacche [204, p. 154] argues that because the savings decision in Conlisk's model endogenously fixes the position of the IPF, the rate of labor augmentation is dependent on economic decisions, even in steady states. Conversely, McCain [324, p. 923] attributes Conlisk's distinct, non-neoclassical results to his "technical progress frontier," defined in terms of *absolute,* not *relative,* increments in the productivity of the factors.

Similarly, Chang's [84, 85] studies of stability appear to show that $\sigma < 1$ is sufficient only for local stability of the Harrod-neutral equilibrium or that σ can take any value without affecting this result, *according to the particular manner in which the IPF is specified.* Thus, "the problem of choosing a particular type of frontier becomes fundamental. It is important to examine under what circumstances it will be correct to choose a particular kind of frontier" [85, p. 211]. We now turn to criticisms of the growth theoretic approach to induced innovation before discussing extensions of the model.

2.2.6. Criticisms of the growth theoretic approach. Kennedy's theory appears to lead to a growth model [122] that incorporates technical change and leaves intact the neoclassical explanations of

[52] However, this result is also contrary to the findings of Nordhaus [366, 368], who does endogenize the rate of change yet concludes that the Harrod-neutral equilibrium is achieved if $\sigma < 1$. See also Kamien and Schwartz [261], who also consider both the rate and direction of technical change in the context of a profit-maximizing firm with a Kennedy frontier. They conclude that the equilibrium is Hicks-neutral (the micro approach is equivalent to no technical change in the capital goods sector) and stable if $\sigma < 1$, unstable otherwise, for both the myopic and dynamic maximization problems.

the steady state, without needing to assume Harrod-neutrality. But the IPF approach raises several other problems. Innovation possibilities must be representable by an IPF of the type suggested by Kennedy and the IPF must be stable. Even then, it cannot explain behavior unless it is known to decision makers [53, p. 37; 557]. Elster [127, p. 105] argues that Kennedy "invokes maximization without a maximizer. He assumes that the innovation will occur at the point on the frontier that, at the ruling factor prices, permits the greatest reduction in unit cost, but he does not tell us how the entrepreneur is supposed to find the frontier and move along it until he finds a maximum, let alone how he is to find the global maximum. The theory lacks microfoundations."

Elster's statement is representative of the views of several critics such as Nordhaus [368] and Samuelson [464], who also question Kennedy's model for replacing exogenous technical change with an exogenously determined innovation possibility frontier.

Ahmad's original critique shows how crucially the results depend on how the frontier is defined. Ahmad [2] argues that the IPF could equally well relate the *amount* of one factor saved per unit of output to the *amount* of the other factor saved. Ferguson [156] shows that the amount saved per unit of output is then $Z = uw + vr$, where u is the amount of labor saved, v is the amount of capital saved, and w and r are factor prices. The IPF is then $u = u(v)$, and cost minimization gives the result $du/dr = -r/w$. Thus factor prices, rather then factor shares, determine the bias, in conjunction with the slope of the frontier. Specifically, if the model is converted to factor shares, then in terms of the factor-augmenting production function of Eq. (1), $u = B(t)L$ and $v = A(t)K$. Thus, $A(t) = v/K$ and $B(t) = u/L$, which implies that the greater the labor input, the lower the level of labor-saving technical change.[53]

Another problem, originally raised by Drandakis and Phelps [123, p. 839] and attributed to Becker, is that a maintained rate of labor augmentation may exhaust the possibilities for further labor augmentation. Not only must the IPF be stable over time in the Kennedy model, but the unit cost reductions, $\hat{A}$ and $\hat{B}$, must be

[53] Hacche [204, pp. 129–132] discusses Ahmad's critique in detail, including the relationships between the IPC and the IPF. There is also a simple derivation of the Harrod-neutral equilibrium.

independent of past increments.[54] This is sufficiently unrealistic that Binswanger [53, p. 38] concluded, "No real world research process can lead to a Kennedy frontier that is independeht of achieved $\hat{A}$ and $\hat{B}$ levels."

To rectify this defect, Nordhaus [368] assumes that there are limiting values of $\hat{A}(t)$ and $\hat{B}(t)$ and that it becomes increasingly difficult to change A and B as these values are approached. Allowing "technological possibilities" to drift over time, he shows that a balanced growth (Harrod-neutral) equilibrium is possible only if the "natural drift" of technology is always Harrod-neutral.[55] This would seem to be equivalent to the original assumption of Harrod-neutral technical change commonly made in simple neoclassical growth models.

More recently the problem of innovation possibilities depletion has been investigated by Magat [307] for the case of the competitive firm (following Kamien and Schwartz [261]). Depletion is allowed by incorporating "depletion factors," u and v, so that the augmentation terms defining the IPF become $vB(t)$ and $wA(t)$. Then, Hicks-neutral technical change occurs only when labor-saving and capital-saving possibilities are depleted at the same rate (since the depletion factors shift the IPF and change its slope). On the assumption that "capital-saving technical advance is easier, or less depleted, than labor-saving technical advance," labor's relative share will increase even at a constant ratio of factor prices. Either this bias in depletion rates, or a falling relative price of capital (due

[54] Compare this with the Evenson and Kislev sampling procedure (Section 2.2.2) in which the payoff from applied research depends on a gap between existing techniques and scientific knowledge. Repeated sampling quickly exhausts the payoff. If this, or a similar view of the research process is accepted, then for the IPF to be stable over time, scientific progress would be required to proceed in such a manner as to replace innovation possibilities at the rate they are "used up." For a discussion of the interaction of science and technology see Rosenberg [438, Ch. 7], who argues that much scientific progess is indeed in response to technical change that has outpaced scientific understanding.

[55] The term *natural drift* is attributable to Samuelson [463, p. 353]. It is important to note that the Nordhaus result applies only to balanced growth equilibria (requiring that the main economic variables remain in the same proportion to each other). The result is damaging because the Kennedy frontier appeared to offer an escape from the assumption of Harrod-neutral technical change. However, for many purposes, a stable equilibrium (that is not on a balanced growth path) is sufficient. Indeed, Magat [307] and Skott [500], discussed below, do derive a stable equilibrium with innovation depletion.

to technical change in the capital goods sector), can explain the rising share of labor observed by some recent authors.

However, neither Magat [307] nor Skott [500], who criticize and extend the model, refers to Nordhaus [368] or attempts to evaluate his contribution. Skott [500, p. 983] argues that in the long run capital augmentation will increase due to "the fact that pure Harrod neutral technical progress would gradually alter the trade-off between the rate of capital augmentation and the rate of labor augmentation." This is true, given the rather arbitrary specification of the depletion factors, but the Magat–Skott model does not follow Nordhaus in considering how change in scientific knowledge determines the payoffs to research activities.

Earlier, Nordhaus [366, pp. 64–54; 367] pointed out that research and development expenditures are assumed to be independent of firm size, which, together with constant returns to factors, ensures decreasing costs and the elimination of competition. He argues that about the only case in which competition can be preserved is when "a new book of blueprints falls from the sky every period"; then the induced innovation model is reduced to "a disguized version of the neoclassical model with exogenous technical change." The alternatives he suggests are that the government must perform the research and transmit it at no cost to competitive firms, or the greater complications of monopolistic behavior must be modeled, leaving little hope of a steady state. This issue of decreasing costs in models that incorporate either research expenditures or learning costs is not a criticism of the Kennedy model alone, but is quite general and frequently overlooked.

A very general cause for complaint is the equilibrium apprach taken by growth theory. The literature on technical change abounds with terms such as *bottlenecks* and *factor scarcity,* which suggest disequilibria. Yet despite Bliss' [62] conclusion that "any interesting technical progress" is incompatible with steady state growth, Robinson [418] is the exception among theorists in considering disequilibria.

2.2.7. Extensions: two-, three- (and more) sector models. The growth-theoretic induced innovation model is extended by Chang [84] to the two-sector case by incorporating separate IPFs within both capital and consumer goods sectors. The basic result is that if the standard capital intensity condition[56] is satisfied and the elasticity of substitution in each sector is less than one, then the Harrod-neutral, steady-state growth path is locally stable (subject to a standard savings assumption). The two-sector, two-IPF model is also investigated by McCain [324], but for the Kennedy-neutral[57] steady state growth path. As in the one-sector model, capital augmentation is zero in the capital goods sector, but in the consumer goods sector the rate of capital augmentation (or disaugmentation) must equal the increase (or decrease) in the price of capital goods. McCain is unable to make a general statement about stability, but Craven [96] proves stability for a simpler Leontief-type model that allows separation of price and quantity equations (and obviously precludes short-term substitutability).

Kennedy [271] generalizes McCain's results to a model with $(m-1)$ capital goods and one consumer good (a crucial assumption), showing that "the rate of factor augmentation in any sector is equal to the rate of change of the price of the factor, if the product of the sector is used as numeraire" (p. 51). This means that although different output and input quantities can grow at different rates, all grow at the same rate in value terms. Commenting on the work of McCain [324] and Kennedy [271], Orosel [383] addresses the problem that if the capital goods are durable, then the continually changing capital goods prices would result in capital gains and losses. These will affect profit rates, which in equilibrium must be the same in all sectors. Orosel proves that there are cases in which capital gains and losses preclude profit rate equalization, but if equalization does occur, the dynamics of the steady state are such that it will be maintained. However, not all concave, differentiable technology frontiers are consistent with the steady-state equilibrium.

Product, as opposed to process innovation, is incorporated in the Kennedy–von Weizsacker model by McCain [325] following

[56] The result depends on the relative shapes of the IPFs, since capital intensities affect factor shares and factor shares are relevant to the determination of technological biases. Following Jones [249], Harrod neutrality in both sectors is imposed as a condition of the steady state path.

[57] "Technical progress is neutral in the sense of Kennedy if the capital-output ratio is constant in *value* terms, in the consumer goods sector" [324, p. 921]. See Burmeister and Dobell [95, pp. 139–146] for a discussion of Kennedy neutrality and a comparison with Harrod neutrality in the context of two-sector models.

Lancaster's characteristic approach to consumer theory. Allowing for the "quality" of goods to be augmented by new product innovation does not lead to fundamentally different results.

McCain [323] and Brewer [69] add a third non-producible factor (land[58] in the simple case, or several fixed resources, in Brewer's extension of the model), thus allowing consideration of the extent to which technical change can stave off the pessimistic forecasts of the classical synthesis. Though the classical stationary state is averted (in the case where population growth is exogenous), an economy with a higher rate of population growth will have a lower rate of growth of income per head. This occurs because in the steady state, the rate of capital augmentation will be zero, while the rate of output growth will equal both the rate of land augmentation and the rate of growth of labor in efficiency units (population growth plus labor augmentation). If population grows more rapidly, land augmentation must increase at the expense of the rate of labor augmentation, which is equal to the growth rate of per capita income.[59]

A third input is also added by Fixler and Ben–Zion [160], but in their model the new input is itself the innovation and the user is a monopolist. The effect on the level of employment of the two other factors depends, not surprisingly, on whether they are substitutable for, or complementary with, the new input. Given how frequently technical change is associated with a new intermediate input (fertilizer in LDC agriculture, for example), it is unfortunate that this case has received so little attention.

---

[58] Fellner [154, p. 1083] raised the issue of the treatment of land, but considered that capital can be substituted with sufficient ease to make the two-factor model applicable. For an informative discussion of fixed resources and technological change, see Rosenberg [434, Chs. 13 and 14].

[59] McCain [323, pp. 498–499] carries the argument further. If population growth is a non-decreasing function of income per capita (in keeping with the classical approach), then the long-run equilibrium is Malthusian, with zero rates of capital and labor augmentation and consequently a constant (subsistence) wage. Brewer points out a problem with disaggregation. In his model, as in McCain [324] and presumably Kennedy's [271] generalization (which does not address the stability issue), the results depend on all $\sigma < 1$. The problem is that "one might reasonably guess that increasing disaggregation to a larger number of factors would make it very likely that at least some elasticities would exceed unity, for example, between two similar types of land in nearby locations" [69, p. 292].

2.2.8. Applications: the environment, utility regulation, and class warfare. McCain [327] provides an application of the Kennedy approach (and the Nordhaus isotech analysis) to environmental policy. In this case the third input is an unpriced collective good, which may be called "environmental capacity"; the model shows that in a growing economy, increasing pollution[60] will only be averted if the "price" (corrective tax, or an alternative measure) of pollution rises at least proportionally with the productivity of labor.

The notion that regulation of the rate of return on capital should lead to overcapitalization is suggested by the static resource allocation model of Averch and Johnson [19]. Following the approach of Kamien and Schwartz [260, 262], Smith [502] incorporates an IPF in the Bailey and Malone [22] model of the regulated firm and shows that rate of return regulation will increase the labor-saving bias of technical change, adding dynamic misallocation of resources to the static inefficiency (an issue raised by Hayami and Ruttan [219, p. 151]. Okuguchi [378] proves that in the CES case, the usual condition of $\sigma > 1$ is necessary for this result, while Magat [306] shows that even in this case the result does not hold generally for the class of homothetic production functions.[61]

A third application of the model to growth cycles is provided by Shah and Desai [489], who incorporate Kennedy's frontier in Goodwin's [180] model of cycles in growth rates. Whereas in Goodwin's model the economy moves in cycles around the equilibrium, giving the capitalists an extra weapon in the form of choice of the bias in technical change leads to a locally stable equilibrium characterized by Harrod-neutral technical change. The authors suggest that the next task of theoretical research should be to model the worker's reaction to the labor-augmenting bias of technical change.

---

[60] Market failure may lead to technical change biased in the direction of environmental pollution. See the critiques of the microeconomic approach in the last section, which emphasize that if the induced innovation model relies on market forces when there are serious distortions, it will lead to suboptimal solutions.

[61] Magat then proves that for both the regulated and nonregulated cases, technical change will be Hicks-neutral and stable if $\sigma < 1$. But for all homothetic functions, $Kr/Lr > Ku/Lu$, where r stands for regulated and u for unregulated. Though Magat does not comment on this point, it would appear to mean that technical change will exactly maintain the initial (static) overcapitalization. Regulation is not confined to utilities. Hayami and Ruttan [219, pp. 151–152] argue that the commodity programs (especially acreage allotments in tobacco) have distorted research resource allocation in U.S. agriculture.

2.3. Estimates of non-neutral technical change and tests of the induced innovation hypothesis

The empirical investigations of the induced innovation hypothesis follow from the discussion of microeconomic approaches in Section 2.2.2. Kennedy's model and the many developments of it represent contributions to the modern theory of economic growth and do not lend themselves readily to empirical tests.[62] By contrast, many contributors to the microeconomic approach were mainly interested in developing a theoretical structure rigorous enough to impose restrictions on the parameters of the models sufficient to allow meaningful empirical tests. Several empirical contributions to the literature on biased technical change are included in Section 2.3.2, since in combination with knowledge of the trend in factor prices they provide useful evidence on induced innovation.

2.3.1. In agriculture. The induced innovation hypothesis was first tested by Hayami and Ruttan [218, 219] against the historical evidence of productivity growth in Japan and the United States for the period 1880–1960. The analysis was extended by Wade [556] to include the United Kingdom, France, and Denmark and by Weber [559] to include Germany. The results for all six countries, with two time periods for Germany and France giving a sample of eight, are fully reported and discussed in Yamada and Ruttan [577, pp. 522–528] and Binswanger and Ruttan [57, pp. 59–86]. The Japan and United States results are updated to 1980 in Hayami and Ruttan [222].

The model of biological technical change in Figure 2.3b suggests that a decline in the price of fertilizer relative to land will induce advances in crop technology, such as fertilizer-responsive crop varieties, characterized by the shift from point *A* on IPC_0 to point *C* on IPC_1. Thus there should be a strong negative relationship between fertilizer per hectare and the price of fertilizer relative to the price of land. Furthermore, there should be a positive relationship between fertilizer per hectare and the price of labor relative to

[62] Binswanger [53, p. 38] attributes the demise of Kennedy's approach to its lack of microfoundations and particularly to the difficulty of conceptualizing an empirical counterpart of the IPF. However, Woodland [574] takes the IPF to be the dual of Diewert's [116] revenue function and constructs an empirical model. Even so, it is the Ahmad/Hayami and Ruttan approach to which most empirical work refers, probably because the theory is a modification of a favorite empirical tool, the production function.

the price of land, since a rising relative price of labor should induce farmers to substitute fertilizer and other chemical inputs such as herbicides and insecticides for more labor-intensive husbandry practices.

The tests[63] regressed the logarithm of the fertilizer/land ratio on the logarithms of the two price ratios, giving results that strongly support the inducement hypothesis.[64] Thirtle [539, p. 175] applies the same test to pooled data for the ten United States farm production regions over the period 1939–78 for wheat, corn, soybeans, and cotton. Again the results are almost entirely consistent with the hypothesis. Wade and Weber also apply the test to the relationship between feed concentrates and factor prices, since in animal agriculture food concentrates play a role analogous to fertilizer in crop agriculture. For the four equations fitted, all eight coefficients have significant signs supportive of the hypothesis.

The mechanical technology model in Figure 2.3a implies that both the land/labor ratio and the ratio of machinery to labor should be negatively related, firstly, to the price of land relative to the price of labor, and secondly, to the price of machinery relative to the price of labor. Tests of the machinery-to-labor relationship produce results that generally support the hypothesis. The coefficient of the machinery price/labor price ratio is always in agreement with the theory, while the coefficient of the price of land relative to the price of labor is contrary to the hypothesis in two of the eight cases.

However, the land/labor ratio equations fail to provide support for the hypothesis. Though the coefficient of the price of machinery relative to the price of labor is contrary to the hypothesis in only one case, the coefficient of the price of land relative to the price of labor fails to support the hypothesis in the majority of cases. Thirtle's [539, pp. 176–177] crop-specific results were clearer; in all eight equations the sign of the machinery price/labor price coefficient is consistent with the hypothesis, but in all cases the sign of the land/labor price ratio is contrary to the predictions of the theory [542].

---

[63] The results reported are from Binswanger and Ruttan [57]. Extending the Japan and U.S. data to 1980 causes two sign changes [222].

[64] See Sahota's [460, pp. 717–728] review article for a methodological critique of these simple tests; he argues that assumptions have replaced implications in this formulation. The functions tested are ad hoc in the sense of not being derived from a specific production relationship.

Two explanations of these perverse results are offered. Ruttan *et al.* [455, pp. 62–64] attribute them to an "innate labor-saving bias" in technological possibilities. Thirtle [542] argues that the results are not damaging to the induced innovation hypothesis but suggest that the Hayami–Ruttan model should be reformulated. Specifically, if machinery replaces land in Figure 2.3a and technical change is depicted as in Ahmad's model [2], Figures 2.3a and 2.3b can be combined to form the four-quadrant diagram shown in Figure 2.4, which allows for interaction between the two groups of factors in Eq. (10). Thus an "innate labor-saving bias" can be explained by the rate of mechanical technical change reducing the input of labor faster than biological innovation reduces the input of land even at constant factor prices and with no direct substitution of land for labor.

These simple tests of induced innovation do not distinguish between technical change and factor substitution. Though this feature of the model appears to have worried some critics, Figures 2.3a and 2.3b suggest that the shorter-run substitution from point *A* to point *B* becomes irrelevant in the secular period in which the IPC shifts and the final equilibrium at *C* is attained.[65] Thirtle [542] applies this reasoning to a four-quadrant model similar to Figure 2.4 but including Hayami and Ruttan's IPCs and changing factor prices. The entire substitution of machinery for labor is attributed to mechanical technical change, just as the whole decrease in the input of land per unit of output is attributed to biological technical change. This allows the labor-saving bias to be measured by the increase in the land-labor ratio.[66] Using estimates for four U.S. field crops, the more initially labor-intensive the crop, the greater the labor-saving bias of technical change over the period 1939–78. This result is entirely consistent with the induced innovation hypothesis.[67]

[65] Hayami and Ruttan [222, pp. 178–187] consider that their tests of the behavior of factor ratios in response to factor price changes relate to movements along the IPC or its production function equivalent, called the metaproduction function.

[66] Net of *direct* substitution of land for labor (in the northwest quadrant), calculated by applying the actual change in the wage rental ratio to estimates of the elasticity of substitution of land for labor.

[67] If the initial production function is labor-intensive, that is, if it requies large amounts of labor relative to capital, expected discounted wage costs will be higher than if the initial production function is capital-intensive. Hence, for a given factor cost ratios and innovation possibilities, labor-saving research is more attractive if one starts from a labor-intensive point than if capital intensity is already high [54, p. 105].

Later tests comply with the neoclassical orthodoxy by differentiating between factor substitution and biased technical change.[68] Thirtle's [540] model is actually simplified by this change, as is shown in Figure 2.4 and explained in the associated discussion. Applied to four U.S. field crops, this model also gives results that entirely support the induced innovation hypothesis. A further test showed that the labor-saving bias was greater for U.S. farm production regions with high ratios of labor to land than for less labor-intensive regions, again supporting the hypothesis.[69]

Ruttan *et al.* [455] compute the elasticities of substitution necessary to explain the observed differences in the land/labor ratios between countries, and the changes over time within countries, for the six-country sample discussed above. In cases where the required elasticity exceeds the econometrically estimated[70] actual elasticity by a sufficiently wide margin, the null hypothesis of neutral technical change is rejected.

The results indicate that four different time paths can be distinguished. Firstly, in 1880 the United States was on the same production function as France, Germany, and the United Kingdom, but after 1880 technical change in the United States had a strong labor-saving bias. Secondly, Continental Europe experienced neutral technical change until the 1960s, after which France and Denmark experienced labor-saving technical change. Thirdly, technical change in the United Kingdom was neutral until 1930 and strongly labor-saving thereafter, though technology remained more labor-intensive than in the United States case. Lastly, Japan began from an extremely labor-intensive position and showed neutral technical change, with a slight labor-saving bias in recent years. These results are largely consistent with the induced innovation

[68] Thirtle [543] also considers returns to scale.

[69] At face value this reasoning is directly opposed to that of Hayami and Ruttan. The two are reconciled by noting that Binswanger is considering a single crop or industry in a developed economy, whereas Hayami and Ruttan have in mind the factor endowment of the entire economy over a long period. Hayami and Ruttan treat the factor intensities as fixed, whereas for Binswanger they are the target variables.

[70] The estimated elasticities are from Binswanger [49]. The data used were for U.S. agriculture 1949–64. These estimates of the elasticity of substitution are used in cross section tests for 1880, 1930, 1960, and 1970 and are applied to all six countries in the time series tests.

hypothesis, but Japan, Britain, France, and Denmark all experienced periods when technical change had a labor-saving bias despite a falling ratio of wages to land prices. This is explained by either an "innate labor-saving bias" or by the international transfer of mechanical technology developed in the United States and not entirely suited to European factor-price ratios.

Some tests described above, like many other recent contributions, use estimates of factor substitution possibilities that are derived from flexible functional forms and that exploit the duality relationships between production, cost, and profit functions. The most popular specification in agricultural economics has been the translog cost function, which gives rise to simple linear systems of factor share equations.[71] Binswanger's [50, 52, 56] many-factor tests follow this course. First, translog cost functions are estimated, using 1949–64 data, to determine the Allen partial elasticities of substitution for inputs of land, labor, machinery, fertilizer, and other inputs. Then the translog function is fitted to time series data for 1912–68 and the changes in factor shares are divided into two elements (using the elasticity measures): the change in factor shares attributable to factor substitution and residuals attributable to technical change. In terms of the factor share definition of bias (Eq. (8)), technical change is found to be fertilizer- and machinery-using over the entire period, with a labor-saving bias discernible after 1948. When plotted against input prices, the technical change indices show trends and turning points consistent with the induced innovation hypothesis (see Ruttan [451, pp. 19–20] for a discussion). In an earlier study, Binswanger [48] derived similar results for the United States since the turn of the century and applied the same approach to Japanese agriculture (see below).

Binswanger's results are confirmed by Chambers and Lee [82], who fit a translog indirect production function to aggregate U.S. data for 1947–80. Technical change is found to be land- and labor-saving and capital- and material-using. These conclusions can be compared to those of Weaver [558], who applied a translog expected profit function to North and South Dakota wheat data for 1950–70. With inputs of labor, capital, fertilizer, petroleum products, and materials, Weaver finds technical change to be labor-

[71] See Wyatt [576, pp. 98–101] for a better explanation and some applications.

saving relative to all other inputs (supporting the results of Lianos [294]), but capital-saving relative to all inputs but labor. Additionally, technical change is fertilizer-using relative to all inputs and petroleum-product-using relative to all inputs except fertilizer.

However, Lopez [299] applies a modified generalized Leontief cost function to time series data for Canadian agriculture[72] over the period 1946–77. He does *not* impose constant returns to scale and finds that the null hypothesis of zero factor-saving technical change cannot be rejected. The constant returns to scale hypothesis, by contrast, is decisively rejected, suggesting that increasing returns are an important source of productivity growth.

Kislev and Peterson [276] are also critical of the inducement hypothesis and have estimated a model of the U.S. agricultural sector that accounts for increases in both land/labor ratios and farm size "by changes in relative factor prices without reference to 'technical change' or 'economies of scale.'" In response, Hayami and Ruttan [222, pp. 187–205] follow Binswanger [56] in developing a framework for decomposing the changes in factor shares into factor substitution and technical change components. Using a two-stage CES, they generate estimates that show both effects to be important.[73] Plots of factor prices against factor-using biases show clear negative relationships for labor, power, fertilizer, and land, thus offering clear support for the inducement hypothesis. The estimates of factor-saving biases are broadly consistent with Binswanger's [51] results except in the case of machinery, where Binswanger found a machinery-using bias combined with a rising relative price. Adjustment of the machinery price series for quality changes removes this inconsistency. Thirtle [541] follows the same methodology and finds biased technical change to be crucial in explaining mechanization in U.S. corn production, while increases in fertilizer per acre can be largely explained by factor substitution. Again the results support the inducement hypothesis with technical change strongly labor-saving, machinery- and fertilizer-using, and neutral with respect to land.

---

[72] In one unusual application, Klein and Kehrberg [278] develop a method of evaluating agricultural research projects based on the innovation hypothesis. They apply it to an existing Canadian animal breeding project.

[73] Kislev and Peterson [275, 276] argues that the mechanisation of agriculture is technical change in the farm machinery industry and factor substitution in agriculture.

For Japan, both the original Hayami and Ruttan tests [218, 219] and the comparisons of Ruttan *et al.* [455] are less convincing than in the U.S. case.[74] Several other studies should also be considered. Sawada [475] surveys the earlier literature and fits a CES function for the period before the First World War, the inter-war period, and the period since the Second World War, finding technical change to be land-saving and labor-using in the first two periods and land-using/labor-saving in the last period.

Binswanger [48] fits a translog cost function to Japanese data since the turn of the century for comparison with similar estimates for the United States. His results also offer some support for the inducement hypothesis. Technical change is found to be land- and machinery-saving. A fertilizer-using bias appears earlier than in the U.S. case, but after the 1920s, technical change is neutral with respect to fertilizer. For labor, the bias is labor-using before 1928 and labor-saving after that date.

The findings of other researchers appear to be contrary to the inducement hypothesis. Yeung and Roe [580] fit a CES function in which the exponential technical change parameters are a function of an index of the price a labor relative to the price of land and find technical change to be labor-saving. This conclusion is also reached by Nghiep [364], whose approach slightly modifies that of Binswanger [49]. Technical change is found to be considerably labor-saving and fertilizer-using for the period 1905–39. Slight machinery-using and land-saving biases are also apparent. However, Nghiep argues that these results *support* the induced innovation hypothesis since agricultural wages rose the most rapidly, followed by the prices of land, machinery, and other inputs, with fertilizer prices rising least.[75]

The studies considered above are based largely on national statistics that have been the subject of debate for the last 20 years

---

[74] The data used by Hayami and Ruttan [218, p. 1117, Table 1] show that the price of labor relative to land was rising in the United States and falling in Japan for the entire period 1880–1960. However, Figure 1, p. 1118, shows a slow rise in the Japanese land/labor ratios, which would account for the "wrong" sign when their simple test was applied to the Japanese data.

[75] Nghiep's price data are clearly different from the series used by Hayami and Ruttan. In particular, the price of land is a rent, rather than a land value. In discussing the partial elasticities of substitution, Nghiep does not mention the oddest result—that land and fertilizer appear to be complements. See also Hunt [235], who suggests that the translog function may be inappropriate in this case and that the parameter estimates may be biased.

[see 222, p. 164, for references]. Kako [256] followed Binswanger's translog cost function approach, but applied it to rice production data for 1953–70 in the Kinki agricultural district. His estimates of technical change are far more consistent with the inducement hypothesis. Innovation saves the scarce factors, labor and land, considerably, but fertilizer and machinery only slightly, thus helping to explain the increasing ratio of machinery to labor and fertilizer to land.

Lee [290] also used different data covering rice production in four prefectures for the period 1955–75 and identified three phases of technological change. From 1957 to 1960, technical change was land-using and labor-saving; from 1961 to 1967, the converse was true; from 1968 to 1975, technical change was again land-using and labor-saving. With respect to machinery, technical change was neutral until 1965 and has shown a remarkable machinery-using bias since that time. Lee argues that these results do support the induced bias hypothesis, though farm size, output price, and lags in innovation and diffusion must also be considered. There is also an anomalous result: technical change has been fertilizer- and pesticide-saving, despite a substantial decline in the relative price of these inputs.

With the benefit of hindsight, Hayami and Ruttan [222] tackle the data problems[76] and exclude the period of the Second World War and its aftermath in an application of their two-stage CES model. For all four inputs (labor, power, land, and fertilizer) their indices show a strong and consistent negative relationship between factor prices and factor-using biases that is entirely consistent with the inducement hypothesis.

The body of evidence is sufficient to substantiate the case for a relationship between factor prices and factor biases. Alderman [7] suggests that factor prices alone are not sufficient to account for the direction of factor biases and proceeds to include the effect on factor shares of research, extension, and infrastructure investment.

The available evidence has been extended to include other countries besides the United States and Japan and the limited information on the United Kingdom, France, Germany, and

[76] See Hayami and Ruttan [222, Ch. 7] and also a paper by Kawagoe, Otsuka and Hayami [266], on which this work is based. Their measurement of labor is in work hours, whereas Nghiep [364] used number of workers.

Denmark [455]. Investigations of non-neutral technical change and/or tests of inducement in agriculture include McKay *et al.* [329] on Australian sheep, corn, and beef production; Ahmad and Kubursi [3] on Egypt and Syria; Park [388] on Korea; Johnson [244] on New Zealand; and Godden [176] on the United Kingdom. The degree of agreement varies, but on balance these studies offer support for the inducement hypothesis. In addition, de Janvry's [109, 111] work on Argentian and Feeny's [143] study of Thailand extend the model to include social and political factors. Sanders and Ruttan [467] investigate the effects of price distortion, and Barlow and Jayasuriya [26] show that new technology in the Malaysian rubber industry tended to be capital-using and hence favored the large estates relative to the smallholders.

Ruttan [451, pp. 20–21] suggests that much could still be learned by extending the tests to include countries that have invested heavily in the green revolution technology and by investigating recent relative price changes, particularly those for energy inputs and agricultural land. However, the basic idea that factor biases are dependent on factor prices is well supported by a large body of evidence. Conceptual advances would seem to be required at this stage. Table I presents the work to date in an accessible form. The studies covered in Table I leave little doubt that there is a relationship between differences or changes in factor prices and the direction of factor-saving bias in technical change. Efforts should be made, however, to design more rigorous tests for induced innovation.

2.3.2. In industry. Few direct tests of the induced innovation hypothesis have been conducted for the non-farm sector, but studies of biased technical change provide a considerable body of supporting evidence that will also be considered.

Using sample survey methods in an early attempt to determine the effect of union wage pressure on technological discovery, Bloom [63, p. 615] found little evidence that union wage pressure led to labor-saving changes, though it did seem "to have produced some increase in the total volume of discoveries." A similar study of the manufacturing sector by Piore [397], based on interviews with engineers, personnel, and industrial relations specialists, reached the same negative conclusion. Enos' [128] study of the petroleum

TABLE Ia

Summary of empirical studies of induced innovation in agriculture—USA (see notes at foot of Table Ic for meaning of symbols)

Study and reference	Crop, area and time period	Methodology	Factor biases	Consistency of results with induced innovations
1) Hayami & Ruttan [219]	National Aggregate, 1880–1960	Regression of factor ratios on factor price ratios	Not estimated	Consistent with induced innovation
2) Hayami & Ruttan [222]	National Aggregate, 1880–1980	Regression of factor ratios on factor price ratios	Not estimated	Consistent with induced innovation but not all coefficients are significantly different from zero
3) Binswanger [51], [56]	National Aggregate, 1912–1968	Translog cost function	Land – (slight) Labour – (after 1948) Machinery + Fertilizer + Other none	Consistent
4) Kislev & Peterson [276]	National Aggregate, 1930–1970	Implicit two stage CES production function	No technical change	Contrary in the sense that factor substitution explains all changes without reference to technical change
5) Weaver [558]	Wheat, North and South Dakota 1950–1970	Translog profit function	Labor – Capital – Fertilizer + Petroleum + Products	Consistent

6) Thirtle [540, 543]	Wheat, Soybeans, Corn and Cotton, 10 US Farm Production Regions, 1939–1978	CES/Cobb-Douglas	Labour-saving bias	Consistent
7) Thirtle [541]	Corn, 10 US Farm Production Regions, 1939–1978	Two stage CES production function	Land + slight Labor + strong Machinery + strong Fertilizer + strong	Consistent
8) Thirtle [542]	Wheat, Soybeans, Corn, Cotton, 10 US Farm Production Regions, 1939–1978	Calculated from the data using 2 stage CES estimates of substitution elasticities	Labor saving bias	Consistent
9) Hayami & Ruttan [222]	National Aggregate, 1880–1980	Two stage CES production function	Land – Labor – Machinery + Fertilizer +	Consistent
10) Chambers & Lee [82]	National Aggregate 1947–1980	Translog indirect production function	Land – Labor – Capital + Materials +	Consistent

TABLE Ib
Summary of empirical studies of induced innovation in agriculture—Japan

Study and reference	Crop, area and time period	Methodology	Factor biases	Consistency of results with induced innovation
1) Sawada [475]	National Aggregates 1883–1963	2 factor CES production function	Land – before 1931 Labor + Land + after 1946 Labor –	Consistent
2) Hayami & Ruttan [219]	National Aggregates 1880–1960	Regression of factor ratios on factor price ratios	Not estimated	Consistent with induced innovation in two cases out of three
3) Hayami & Ruttan [222]	National Aggregates 1880–1980	Regression of factor ratios on factor price ratios	Not estimated	Consistent
4) Binswanger [48]	National Aggregates 1900–1970	Translog cost function	Land – Labour + before 1928 – after *Machinery + Fertilizer +	Mainly consistent
5) Yeung and Roe [580]	National Aggregates 1880–1940	Modified 2 factor CES	*Land + *Labor –	Contrary to induced innovation

6) Kako [256]	Rice in Kinki District 1953–70	Translog cost function	Land – strong Labor – strong Machinery – slight Fertilizer – slight Other – very slight	Consistent
7) Nghiep [364]	Rice, national, 1905–39	Translog cost function	Land *Labor – Machinery + Fertilizer +	Mainly consistent
8) Lee [290]	Rice, 4 prefectures 1955–77	Translog production function	Land + 2 out of Labor – 3 sub-periods Machinery – strong *Fertilizer – * & Pesticide –	Mainly consistent
9) Alderman [7]	National Aggregates 1902–1940	Translog cost function	Land – *Labor + Capital + Current inputs + (Mainly Fertilizer)	Mainly consistent
10) Hayami & Ruttan [222]	National Aggregates 1880–1980	Two stage CES production function	Land not clear Labor – Machinery + Fertilizer +	Consistent

TABLE Ic
Summary of empirical studies of induced innovation in agriculture—other countries

Study and reference	Country	Crop area and time period	Methodology	Factor biases	Consistency of results with induced innovation
1) McKay, Lawrence & Vlastuin [329]	Australia	Sheep and wool, crops, beef cattle, and other farm output in the Australian wheat/sheep zone, 1952–1977	Translog variable profit function	Land – slight Labor – strong Capital + strong *Materials – slight	Mainly consistent
2) Lopez [299]	Canada	National Aggregates, 1946–1977	Generalized Leontief cost function	Increasing returns to scale and no technical change	Contrary
3) Ruttan *et al.* [455]	Denmark	National Aggregates, 1910–1965	Regression of factor ratios on factor price ratios	Not estimated	Mainly consistent, but the behaviour of the land/labor ratio in Denmark, France and the U.K. is not explained by the relative factor prices
	France	National Aggregates, 1870–1965	Regression of factor ratios on factor price ratios	Not estimated	
	Germany	National Aggregates, 1880–1968	Regression of factor ratios on factor price ratios	Not estimated	
	United Kingdom	National Aggregates, 1870–1965	Regression of factor ratios on factor price ratios	Not estimated	
4) Ruttan *et al.* [455]	Germany	Livestock Production 1880–1968	Regression of factor ratios on factor price ratios	Not estimated	Consistent

	Denmark	Livestock Production 1880–1925	Regression of factor ratios on factor price ratios	Not estimated	
	United Kingdom	Livestock Production 1870–1965	Regression of factor ratios on factor price ratios	Not estimated	
5) Binswanger [52]	USA	National Aggregates, 1880–1970	International comparisons of residual measures of land and labor-saving technical changes	Labor-saving	General support
	Japan	National Aggregates, 1880–1970		Mainly neutral	
	Great Britain	National Aggregates, 1880–1970		Labor-saving	
	France	National Aggregates, 1880–1970		Slightly labor-saving	
	Germany	National Aggregates, 1880–1970		Slightly labor-saving	
	Denmark	National Aggregates, 1880–1970		Slightly labor-saving	
6) Park [388]	South Korea	Rice, 1963–1984	Regression of factor ratios on factor price ratios	Not estimated	Consistent
7) Barlow and Jayasuriya [26]	Malaysia	Rubber, 1922–1978	Cost calculations	Land – Labor – Capital +	Consistent, but a bias in favor of plantations relative to small-holders
8) Johnson [244]	New Zealand	National Aggregates, 1945–1967	CES production function	Labor – Capital +	Consistent
9) Godden [176]	United Kingdom	National Aggregates, 1950–1980	General linear variable profit function	Land – Labor – No Result Plant + *Materials –	Unclear, generally poor results

For attempts to examine the social and political aspects of induced innovation see de Janvry [96], [97], Feeny [127], Sanders and Ruttan [394] and de Janvry and Dethier (1985).

NOTES For factor biases (–) indicates factor-saving technical change and (+) indicates that technical change was factor-using. The symbol * against a factor bias indicates that it is either contrary to the direction indicated by the inducement hypothesis or explanation by the author is required to reconcile it with the theory.

industry found that inventions tended to be neutral, rather than saving the scarce factor, though there was some labor-saving bias at the development and improvement stage.

Early confirmation of non-neutral technical change in the United States was provided by studies such as Brown and Popkin [73], Resek [416], and Brown [72]. David and Van de Klundert [103] found that the U.S. private economy exhibited a significant Hicksian labor-saving bias during the first half of this century. This result is confirmed by the more recent work of Sato [468], Takayama [531], Panik [387], and Zind [581]. At the industry level, Gupta and Taher [199] find the same labor-saving bias for post-war cotton textile production in the United States, as do Bergstrom and Melander [37] for nine Swedish manufacturing industries and Forsund and Jansen [161] for the Norwegian aluminum industry.

Subdividing the time period. David and Van de Klundert [103] found technical change in the U.S. private sector to be labor-saving for the 1900–18 period at a rate greater than for the full time span, neutral from 1919 to 1945, and labor-saving at a still greater rate from 1946 to 1960. However, Brown and Popkin [73] divided the period into three "technological epochs," 1890–1918, 1919–37, and 1938–58, and found technical change to be labor-saving between the first pair of epochs, but capital-saving between the second pair.[77]

The results of Morishima and Saito's [343] test of induced innovation partially agreed with both earlier studies, finding labor-saving change over the entire period, but with a heavy labor-saving bias before 1929 and a slight capital-saving bias thereafter. To test the induced innovation hypothesis, total technical change was divided into an induced component, an autonomous component, and the effect of changing industrial composition. Though the 1902–29 period was dominated by the growth of the industrial sector relative to agriculture, induced innovation was found to be capital-saving in the depression years of 1929–38,[78] when the labor force was increasing relative to the capital stock, and labor-saving for the period of high employment and rapidly rising wages from 1938 to 1955.

---

[77] A more complete survey of early two-input studies is provided by Kennedy and Thirlwall [292].

[78] This is supported by Uselding and Juba [546], who found total technical change to be capital-saving during the decade of the 1930s.

Fellner [155] also set out to investigate the induced innovation hypothesis, using U.S. data and arguing that increases in the capital-labor ratio would have increased labor's share of income but for offsetting effects. His regression results suggest that labor-saving innovations were one of these opposing forces.

The recent introduction of flexible functional forms that can sensibly accommodate several inputs, together with the considerable change in relative factor prices caused by the energy crisis, has given a new lease on life to studies of non-neutral technical change. Berndt and Khaled [40] fit a generalized Box–Cox cost function to U.S. manufacturing data for the period 1947–71. Technical change is found to be capital- and energy-using and labor- and intermediate-material-saving. Woodward [575] applies Binswanger's [56] methodology to the postwar U.S. manufacturing sector and finds that labor augmentation is most pronounced, followed by capital augmentation, while the trends for energy and materials are far less clear. These results are not at all contrary to the inducement hypothesis, but those of Jorgenson and Fraumeni [252] are. They estimate biases of technical change for 35 U.S. industries. Technical change is labor-using for 31 industries, energy-using for 29, capital-using for 25, and material-saving for 33 of the 35 industries.

Wills' [569] four-factor test of the inducement hypothesis, using data on the U.S. primary metals industry, is entirely supportive of the hypothesis. The rate of augmentation in descending order is labor, energy, materials, and capital, which corresponds exactly to the ordering of factor price increases. Moroney and Trapani's [345] four-factor study of six U.S. natural-resource-using industries also produces results consistent with inducement. Berndt and Wood's [41] study of electric power produces a result similar to that of Wills [569]. Several other contributions on electricity generation reach similar conclusions [98, 178, 186], finding technical change to be capital- and labor-saving and fuel-using. Investigating the same industry, Stevenson [514] finds technical change to be capital- and labor-saving and fuel-using but argues that his "results failed to demonstrate the existence of induced technological bias" [514, p. 172]. Belinfante's [32] productivity study of U.S. electricity generation is also negative, finding evidence of technical change but little sign of bias.

For other countries the evidence is more limited, but Norsworthy and Malmquist's [369] study of the productivity slowdown is based on the translog production function and compares estimates of biased technical change in U.S. and Japanese manufacturing. For Japan, technical change is capital-using and labor-, energy-, and material-saving. The strong energy-saving bias for Japan contrasts with energy-using technical change in the United States and is attributed to high Japanese oil prices. Investigating the Japanese petrochemical industry, Lau and Tamura [286] apply a modified Leontief production function but cannot reject the hypothesis of zero technical change.

Rao and Preston [407] provide estimates of the factor-saving biases in many Canadian industries. For the majority, technical change is labor-saving, and capital-, energy-, and raw-material-using. Duncan and Binswanger [124] compare estimated rates of factor augmentation with rates of factor price change for data on Australian manufacturing industries and claim mild support for the inducement hypothesis. However, their translog study [125] of alternative sources of energy for five Australian industries over the period 1948–67 reaches negative conclusions. The price of fuel oil has fallen most relative to the price of other sources, yet oil expenditure shares are mainly neutral or factor-using. The price of coal gas has fallen least, but only one industry shows a reduction in the factor share.[79]

For the developing countries there have been few tests, but Lynk [301] studies a range of Indian manufacturing industries for the 1952–71 period. Technical change is found to be labor-saving and plant- and machinery-using. Levy's [293] productivity study of Iraq over the period 1961–67 also shows technical change to be labor-saving. However, he points out that this result *is* consistent with the induced innovation hypothesis, since wages were rising over the period and Iraq is neither labor-abundant nor short of capital, having oil resources and foreign Arab funds.

To date, the evidence of induced innovation in industry rests on a few deliberately constructed tests and a considerable body of

---

[79] The authors point out that technical change did appear to be coal-saving and electricity-using, suggesting that if external costs such as pollution were included the test results might better fit the inducement hypothesis.

information on the direction of technical change, which is sometimes explained by reference to input price movements. Table II provides an overall impression but is no substitute for a careful inspection of the available evidence. Although the weight of evidence from the industrial sector tends to support the inducement hypothesis, the evidence is less clear than from the studies in the agricultural sector. In part this is because many studies have been conducted within a simple two-factor framework. The lack of clarity may also be due to differences in the nature of technical possibilities among industries.

2.3.3. In history. The Rothbarth–Habakkuk thesis, discussed in Section 2.2.1, suggests that the high wage rate in the United States relative to Britain should have resulted in a greater labor-saving bias to technical change in the United States. This proposition was subjected to an empirical test by Asher [17], who fitted a CES function to U.S. and British textile data for the second half of the nineteenth century. Asher [17, p. 440] interpreted the results as showing a labor-saving bias in both countries, but with Britain having the greater labor-saving bias, clearly contradicting the hypothesis.[80]

Further evidence on the United States from a similar study of manufacturing industries over the period 1839–99 by Uselding and Juba [546] shows technical change to be labor-saving over the whole sample (in keeping with the rise in the wage/rental ratio), but finds a capital-saving bias for the decades of the 1840s, 1870s, and 1890s.

Uselding [545] analyses data from the Springfield Armory for 1820–50 and finds technical change to be labor-saving over the entire period (over which time the wage/rental ratio was rising). Division into subperiods shows that the labor-saving result held only for the 1841–50 period, when the relative wage was rising most rapidly. In addition, three-factor analysis shows some support for the Ames and Rosenberg hypothesis, since raw and intermediate

[80] David [100, Ch. 1] shows that Asher's measure of bias ($B = \lambda - \gamma$, where λ is the labor augmentation parameter and γ that for capital) depends on the rates of technical progress. He argues that a *relative* measure of the bias, $B^* = (\lambda - \gamma)/\lambda$, would be more appropriate. However, inspection of Asher's results [17, pp. 439–440] indicates that the rate of technical change in Britain would have had to be practically double the U.S. rate for cotton and triple the U.S. rate for wool for the conclusions to be reversed.

TABLE IIa
Summary of empirical studies of induced innovation in industry

Study and reference	Country, Industry and time period	Methodology	Factor biases	Consistency of results with induced innovation
1) Brown and Popkin [73]	US non-farm domestic sector, 1890–1958	Cobb-Douglas production function parameter estimates are used in comparisons of "technological epochs"	1890–1918, 1919–1937 Labor-saving; 1919–1937, 1938–1958 Capital-saving	The labor-using result is clearly contrary to the induced innovation hypothesis.
2) Resek [416]	US non-farm domestic sector, 1919–1959	Regression of factor ratio on factor prices	Labor-saving	Consistent
3) David and Van de Klundert [103]	US private domestic economy, 1899–1960	CES production function	Labor-saving	Consistent
4) Ferguson [156]	Several US industries, 1949–1961	CES production function	Generally labor-saving	Mainly consistent
5) Sato [468]	US private non-farm sector, 1909–1960	CES and CEDD (constant elasticity of derived demand) production functions	Labor-saving	Consistent
6) Takayama [531]	US non-agricultural private sector 1909–1960	Linear regression of several "simple formulas"	Labor-saving	Consistent
7) Panik [387]	US private domestic sector 1929–1966	CES production function	Labor-saving	Consistent
8) Zind [581]	US private non-farm sector 1909–1960	Linear regression estimation of coefficients, of underlying relationships, similar to Takayama [444]	Labor-saving	Consistent
9) Gupta and Taher [199]	US textile industry 1949–1974	Translog cost function	Labor-saving	Consistent
10) Bergstrom and Melander [37]	Swedish Industry in aggregate and eight industries 1950–1973	Input demand functions from a CES	Labor-saving	Consistent

11) Forsund and Jansen [161]	Norwegian Aluminium industry 1966–1978	Comparison of cost-minimizing factor ratios, as in Salter [390]	Labor-saving	Consistent
12) Morishima and Saito [344]†	US domestic economy 1902–1955	Differentiates between autonomous and induced innovation and includes structural change. A proper test of induced innovation based on the CES	*Induced innovation* is: Practically neutral 1902–1929 Capital-saving 1929–1938 Labor-saving 1938–1955	Consistent
13) Fellner [155]†	US two-digit manufacturing industries 1948–1957	Regression of increase in factor ratio on the rate of change of the factor price ratio	Suggests labor-saving innovations	Consistent
14) Berndt and Wood [41]	US manufacturing, 1947–1971	Translog cost function	Capital Labor Energy Materials: Could not reject Hicks' neutrality	Inconsistent
15) Belinfante [32]	460 observation on 80 US steam-electric generating plants of 1947–59 vintage	Total factor productivity study based on Divisia indices	Little evidence of biases	Inconsistent
16) Wills [569]†	US primary metals industry, 1948–1974	Translog cost function	Labor – Energy – Materials – Capital +	Consistent. The rates of factor augmentation are shown in descending order; the ordering of the rates of price increase is the same.
17) Berndt and Khaled [40]	US manufacturing, 1947–1971	Generalized Box–Cox cost function	Capital + Labour – *Energy + Materials –	Mainly consistent
18) Stevenson [514]†	US electricity generating firms for 1964 and 1974	Translog cost function	*Capital – Labor – Fuel +	The author finds no evidence to support the hypothesis but see (23).

TABLE IIa
(*Continued*)

Study and reference	Country, Industry and time period	Methodology	Factor biases	Consistency of results with induced innovation
19) Jorgenson and Fraumeni [252]	Thirty-five US industries, 1958–1974	Translog cost function	Capital + (25) industries *Labor + (31) industries Energy + (29) industries Materials – (33) industries	Contrary
20) Moroney and Trapani [345]	Six US natural resource intensive industries 1955–1974	Translog cost function	Capital + (3) industries Labor – (6) industries Natural resources + (5) industries Scrap metal + (2) industries (Scrap was used in only 2 industries)	The dominant pattern of labor-saving and natural resource-using technical change is consistent with relative input price movements.
21) Woodward [575]	US non-farm private sector, 1948–1978 US manufacturing sector, 1958–1977	Translog cost function	Capital – strong Labor – strong Energy – weak Materials – weak	Mainly consistent. Energy prices have a slight effect.
22) Greene [186]	US electric power companies 1955–1975	Translog cost function	Capital – Labor – (since 1965) Fuel +	Unclear. The prices of capital and labor were rising relative to that of fuel almost to the end of the period.

23) Gollup and Roberts [178]	US electric power industry 1973–1979	Translog cost function	Capital – Labor – Fuel +	Unclear.
24) Lau and Tamura [286]	20 Japanese petro-chemical plants of vintage 1958–1969	Input demand functions from a modified Leontief production function	Capital, Labor, Energy, Materials: Hypothesis of zero technical change could not be rejected	Contrary
25) Duncan and Binswanger [124]†	14 Australian manufacturing industries 1948–1967	Translog cost function	7 factors	Mild support, but results unclear
26) Duncan and Binswanger [25]†	Energy use in Australian manufacturing industries 1948–1967	Translog cost function	Coal – Fuel oil unclear Electricity + Coal gas unclear	Perhaps mild support
27) Levy [293]	Iraq, industrial sector, 1961–1967	Translog profit function	Labor-saving	Not really contrary. Wages were rising and Iraq is not short of capital
28) Lynk [301]	15 Indian industries, 1952–1971	Generalized Leontief cost function	Capital + Plant + Labor –	Contrary
29) Norsworthy and Malmquist [369]	Japanese manufacturing 1965–1978	Translog cost function	Capital + Labor – Energy – Materials –	Unclear
30) Rao and Preston [407]	Canadian industries 1957–1979	Translog cost function	Predominantly Capital – *Labor neutral Energy + Materials +	Contrary
31) Daly and Someshwar Rao [98]	Ontario Hydro, 1967–1980	Translog cost function	Capital – Labor – Energy + Materials +	Mainly consistent

TABLE IIb

Summary of empirical studies of induced innovation in economic history

Study and reference	Country, Industry and time period	Methodology	Factor biases	Consistency of results with induced innovation
1) Asher [17]	British and American textiles, 19th century	CES production function	Labor-saving bias in both countries	Contrary to the hypothesis since Britain had the greater labor-saving bias.
2) Uselding [545]	Springfield Armory 1820–1850	CES production function	Labor-saving bias overall	Weak support
3) Uselding and Juba [546]	American manufacturing 1839–1899	CES production function	Labor-saving bias overall	Weak support
4) Smith [503]	Springfield Armory 1820–1850	Translog cost function	Labor – Insignificant Capital + Materials –	Unclear. The author avoids making too much of the results
5) Cain and Paterson [78]	19 US manufacturing industries 1820–1919	Translog cost function	Labor – (14) industries Capital + (16) industries Natural resources + (12) industries Other – (13) industries	Mainly consistent
6) Phillips [394]†	British coal industry, second half of 19th century	Translog cost function	Labor-saving bias except during the depression	Consistent
7) Phillips [395]†	Pig iron, cotton textiles, second half of 19th century	Translog cost function	Unclear	Contrary. Nearly all parameter estimates have the wrong sign.

NOTES Studies (1)–(12) are two-factor estimates of non-neutral technical change, rather than tests of induced innovation. Papers that explicitly address themselves to testing the induced innovation hypothesis are marked †.

The symbols (+), (−) and * have the same meaning as in Table I. In most instances "consistent" means only that the direction of the factor bias of technical change is the opposite of relative input price changes. The degree of correlation between the two variables may be high or low.

material inputs appear to be important. This result is confirmed in the critique by Klingaman, Vedder, and Gallaway [279] of Uselding's approach, and Smith's [503] study, which applies a translog function to Uselding's data. Natural resource inputs were found not to be separable (defined in Section 2.2.2) from inputs of capital and labor, suggesting, in corroboration with the Ames and Rosenberg hypothesis, that natural resources must be considered along with labor and capital in evaluating U.S. technical change.[81]

Cain and Paterson [78] fit a translog cost function to data for a large range of U.S. industries over the period 1850–1919. Their results appear to support both the inducement hypothesis and the Ames and Rosenberg position. Over the period, the price of labor rose relative to that of both capital and raw materials. For the majority of industries, technical change was labor-saving and capital- and resource-using (though the results did not necessarily coincide). An individual industry could, for example, be both labor-neutral and capital-using.

The empirical evidence is less clear for the United Kingdom. Phillips [395] has investigated pig iron, cotton textiles, and coal mining [394] for the second half of the nineteenth century. No evidence of induced innovation can be found for pig iron. For cotton, technical change is labor-using during the period of the cotton famine (1854–72), but results for the other periods contradict the inducement hypothesis. For coal, induced innovation is evident except in the 1880s and early 1890s, the period once known as the "Great Depression."

2.4. Alternatives to the conventional approach

Dissatisfaction with the neoclassical approach to factor substitution and technical change was discussed in Section 2.1 above, but this survey shows the range of difficulties involved in analysing technical change. Partly, progress has been slow because of the breadth of the area, but technical change also raises problems such as market failure, interdependencies, historically contingent events, and the dynamics of change, which do not fit easily into the neoclassical

[81] Smith appears to reject factor factor-price-induced, biased technical change. His results indicate that labor-saving technical change was not significant, but that there is significant capital-using and natural-resource-saving technical change.

framework. Nelson and Winter [360, p. 205] have been prominent among critics of the conventional methodology: "But what we know about technical change should not be comforting to an economist who has been holding the hypothesis that technical change can be easily accommodated within an augmented neoclassical model. Nor can the problem here be brushed aside as involving a phenomenon that is 'small' relative to those that are handled well by the theory; rather it relates to a phenomenon that all analysts (or virtually all) acknowledge as the central one in economic growth. The tail now wags the dog. And the dog does not fit the tail very well. The neoclassical approach to growth theory has taken us down a smooth road to a dead end."

If the neoclassical paradigm does prove to be a degenerate research programme in the area of technical change, the causes of the failure lie in its origins. Founded on classical physics and using mechanical analogues, the fundamental concept of neoclassical economics is that of equilibrium, the position to which the spring must return or the pendulum settle to rest. Variety is an unnatural state; the norm will prevail once the perturbations cease to disguise it. To neoclassical analysis, path dependence is an unfortunate complication; it is a "system of thought which in its pure form happens to be fundamentally ahistorical, if not actually anti-historical" [100, p. 11]. If time is dealt with, mathematical tractability is enhanced by seeking out stationarity, which definitionally makes history irrelevant. The failure to have a mechanism for explaining variety and the failure to come to grips with historically contingent events are at odds with the reality of technical change at the micro level. Firms differ, particularly in their technological characteristics, because they have different histories and different past experiences.

By contrast, the "neo-Schumpeterian approach is concerned above all with the process of economic change, as opposed to the analysis of equilibrium states" [163, p. 609]. "Economic progress, for Schumpeter, did not consist of price cutting among harness makers. The competitive behavior that really mattered in the long run came from the innovative acts of automobile manufacturers which abolished harness making as an economic activity" [434, p. 5]. Schumpeter [485, p. 64] himself describes the irrelevance of equilibrium analysis eloquently: "What we are about to consider is that kind of change arising from *within the system* which so displaces its equilibrium point that the new one cannot be reached from the old one by infinitesimal steps. Add successively as many mail coaches as you please, you will never get a railway thereby."

Nelson and Winter [360] have produced an alternative, evolutionary theory of technical change that is not ahistorical[82] and avoids the neoclassical distinction between movements along a production function and shifting the production relationship. The neoclassical constructs of rational maximization and equilibrium are replaced by local search for, and selection of, techniques based on satisficing behavior. Thus the model's intellectual heritage can be attributed to both Schumpeter and the behavioralist approach of Simon [498].

It is particularly pertinent to this survey that even the early versions [356, 357, 362] of the model incorporate a simple price-inducement mechanism that can produce biased technical change in computer simulations. Firms produce with fixed proportion techniques that are retained if profitability is satisfactory, but if profits fall below the critical level, they search for new techniques (or imitate other firms) with a greater probability of finding techniques close to the original production point. In this sense the model is not ahistorical. New techniques, randomly selected, are tested for profitability and accepted if satisfactory. At a high wage/rental ratio, the probability of labor-saving techniques being accepted is greater than at a low wage/rental ratio.

Excess profits are invested, so that the growth of the capital stock is determined by the firm's total investments, with successful firms having the highest weights. The labor supply is inelastic, and firms begin from the same situation. The level of output, the wage/rental ratio, and capital accumulation rates are determined endogenously. Though firms will produce with different techniques, a higher wage rate will favor the choice of capital-intensive techniques, as described above, and lead to the expansion of capital-intensive firms relative to those with a lower capital/labor ratio. Thus, although the search for new techniques is random, in the aggregate the capital/labor ratio will increase if the wage/rental ratio rises.

---

[82] David [100] argues that the Nelson and Winter approach is not historical, but this reasoning has been questioned by Elster [127, pp. 156–157], who also provides a brief introduction to evolutionary theories. His illustrations cover animal tool behavior and the technology of fishing boats. On the methodological foundations of the evolutionary approach to economics, see Winter [572, 573].

Later models [357, 358] retain the uncertainty surrounding the research process, but explicitly introduce ongoing direct research (not dependent on inadequate profitability),[83] which adds a further inducement mechanism. At higher wage/rental ratios, the firm has an incentive to devote a higher percentage of its research effort to sampling the spectrum of capital-intensive techniques.[84]

Some of the computer simulation runs produce parameter values and time paths that appear to "explain" Solow's [506] historical data for the United States as well as the neoclassical analysis does. Indeed, neoclassical explanations or simulations explain the data equally well, leading Nelson, Winter, and Schutte [362, p. 117] to suggest an identification problem. Different theoretical structures can lead to similar statistical patterns so that "a world without a production function can, for example, mimic much of the behavior of a world that has one."

Following Atkinson and Stiglitz [18] (discussed in Section 2.1 above), David [100] has developed an evolutionary model of technical change to investigate the labor scarcity hypothesis of Rothbarth and Habakkuk.[85] He argues that if the abundance of land did lead to an initially high wage rental ratio in the United States, then learning by doing would induce "locally neutral" technical progress, improving the capital-intensive techniques so that switching back to a more labor-intensive technique would not occur even if the factor price ratio changed. Thus, "there is some theoretical basis for seeking the origins of the modern configurations of a society's technology in the accidents of its remote factor-price history" [100, pp. 66–67].[86] Also, following Ames and Rosenberg [11], he argues that if the abundance of raw materials and prodigality in their use did foster more mechanized techniques in the United States than in Britain, the United States would indeed have initially followed a more capital-intensive path even if the

[83] Binswanger [53, p. 32] criticized the assumption that search only begins when profits are unsatisfactory. This assumption is contrary to Schmookler's [481] evidence that increasing demand, and hence higher profitability, leads to more innovation rather than less.

[84] More recent developments of the model, such as by Nelson and Winter [359], concentrate on Schumpeterian hypotheses rather than induced innovation.

[85] See Williamson [568] for a useful critique of David's approach.

[86] Rothbarth [439, p. 387] argued that once established, American superiority became self-reinforcing and operated independently of its original historical cause. David's analysis provides a theoretical basis for this claim.

wage/rental ratio were the same in both countries. The historical tests of the last section should be interpreted in light of this observation.

Radner [405] has developed a behavioral model that combines features of the Kennedy and Nelson and Winter approaches. At each point in time the satisficing manager's behavior, called "putting out fires," requires allocation of his efforts to reduce the quantity of whichever input promises the largest expected cost reduction. Since this expectation depends on factor prices, the model contains an inducement mechanism and is an improvement on the Kennedy model, since innovation depends upon a real resource-using activity, even though the "budget" is fixed at one manager.

An increasing number of authors have now followed Nelson and Winter's lead in either recommending or contributing to the evolutionary or neo-Schumpeterian approach to technical change. See, for example, Elster's [127] methodological study, Metcalfe [334], Frausman [163], Kelly and Kransberg [267], Winter [571], and Iwai [238, 239]. But Nelson and Winter's [360] pioneering study shows that substantial difficulties will be encountered in pursuing the biological analogue. To begin with, the firm has some control over its own destiny that is lacking in an organism, and considerations of this nature prevent the simplistic imposition of biological notions in economics. It remains to be seen if their brilliant contribution will continue to develop and displace the neoclassical approach from its niche, or prove to be yet another dead end.[87]

3.0. THE ADOPTION AND DIFFUSION OF INNOVATIONS[88]

Parts 1 and 2 of this paper considered the process of innovation and the rate and direction of technical change. However, the term

---

[87] The biological concept of niches has been added to the economics literature by Mark, Chapman, and Gibson [321].

[88] This part owes much to two recent studies of diffusion by Davies [105] and Stoneman [522]. Similarly, the discussion of adoption draws heavily on a survey of agricultural adoption by Feder, Just, and Zilberman [138]. It is beyond the capabilities of the authors to do justice to the voluminous literature on diffusion from all the contributing disciplines. We have borrowed freely, since many important developments occurred in other subject areas, but the paper is written for economists. A guide to work in other disciplines is provided by the extensive bibliography in Rogers [423].

technical change is used to refer both to changes in the level of technology itself (often called technological change) and to the *effects* of those changes as they are reflected in productivity increases and the rate of economic growth [272]. Since advances in knowledge are inherently difficult to quantify, most empirical investigations follow the second route and concentrate on the effects of technical change on output levels.[89] Though innovation may determine the best practice technique, the speed of imitation or diffusion of the new knowledge will thus play a major role in determining the measured rate of technical change.[90] Indeed, whereas macroeconomic studies of technical change generally imply immediate diffusion of new technology, recent works on the "productivity slowdown" tend to show that other factors, including a decline in the rate of spread of technology, have been more important than the fall in R & D expenditures [504].

Part 3 of this paper concentrates on explaining the diffusion process, beginning with the best-known simple model, then proceeding to show why and how the theory has been developed. *Diffusion studies* do not consider the innovation process, but begin at a point in time when the innovation is already in use.[91] The earliest adopters may be called innovators, and the diffusion process is the spread of the new technique across the rest of the population. *Adoption studies* consider the reasons for adoption at one point in time, or the reasons for time of adoption for individual users. In contrast, most diffusion models are dynamic and study the behavior of the diffusion process over time. Thus, relative to adoption, *diffusion* may be viewed as a dynamic, aggregative process, over continuous time.[92] Alternatively, "if one can explain the date of

[89] Though this approach may capture the more direct effects of technical change, the important and wide-ranging social, organizational, and institutional changes that can be caused by major technical innovations tend to be neglected. Kuznets [283] argues that these resultant changes would have to be identified and taken account of if we are to calculate the *net* contribution of innovation to economic growth.

[90] The relationships between best practice techniques, diffusion, and productivity growth have been investigated by Shen [495, 496].

[91] This is arguably a shortcoming; Rogers [423, Ch. 4] states that "events and decisions occurring previous to this point have a considerable influence upon the diffusion process."

[92] In statistical terms, the S-shaped "diffusion curve" represents a cumulative distribution, obtained by *integrating* a unimodal frequency distribution of adopters arranged on a time scale. See Stoneman [522, pp. 96–97] and Figures 3.1a and 3.1b.

adoption by individual firms, then by aggregation one should have the inter-firm or intrasectoral diffusion curve" [522, p. 95].[93]

Diffusion research has been multi-disciplinary, as is shown by the historical account provided by Rogers [423] and by the heterogeneous list of references provided here. Indeed, since interest in diffusion began at the turn of the century [532], over 3,000 publications have appeared, with sociology, communications, education, marketing, public health, and geography all accounting for a greater proportion of the literature than does economics. An early review of the several traditions in the study of the diffusion process is provided by Katz, Hamilton, and Levin [265].

For simplicity of exposition, this part continues (Section 3.1) with a description of the "epidemic" diffusion model that has served as a basic research tool in most disciplines, including economics. This is followed (Section 3.2) by an account of the application of the epidemic model to the diffusion of techniques in economics. The procedure has been applied to diffusion within individual firms (intra-firm), between firms within an industry (interfirm, or intra-industry, or intra-sectoral diffusion, also referred to as the rate of imitation), on an economy-wide level, and internationally.

Evaluation and development of the model follows, concentrating on the issues of interest to economists. Thus, Section 3.3 considers the manner in which different diffusion curves are generated when the stringent assumptions of the simple epidemic model are altered. Section 3.4 reviews adoption studies that explain why individual firms are leaders or laggards in the use of new techniques. Section 3.5 extends the analysis by considering recent attempts (threshold models and game theory) to provide a sound theoretical basis for the diffusion process. The next part considers the *supply* of

[93] This statement does disguise the fact that adoption and diffusion studies are examples of two distinct modes of explanation in social science research. Mohr [340] argues that adoption studies are an example of *variance theory,* in which the independent variables are simultaneous, necessary, and sufficient to explain the variance of the dependent variable. This approach, common in neoclassical economics, is contrasted with *process theories,* such as diffusion, that predominate in the other social sciences. In process theory, which deals with discrete states and events, the precursors are necessary for the outcome to occur, but are not sufficient, since the process is probabilistic and the outcome only follows a particular time sequence of precursors. See Section 3.4 of this part, where adoption studies are discussed, and the Conclusion, which considers methodoligical issues.

innovations. In Section 3.7, aspects of the international transfer of technology are briefly considered. Finally, the Conclusion takes stock of the current state of knowledge, particularly from a methodological perspective.

3.1. The epidemic model

If personal contact is important in the adoption of an innovation by a limited population, the diffusion process may be viewed as formally akin to the spread of an infectious disease [16, p. 33]. One simple form of the epidemic model may be described by the differential equation,

$$\frac{dn_t}{dt} = \beta \frac{n_t}{N}(N - n_t) \tag{17}$$

where n_t is the number of individuals who have contracted the disease (adopted the innovation) at time t, N is the fixed population (of potential adopters), and β is the parameter reflecting the likelihood of contracting the disease. Thus the number of new infections (adoptions) at period t is equal to the number of uninfected persons (remaining potential adopters), $N - n_t$, multiplied by the probability of infection (adoption), which is the product of the proportion of the population infected (already adopters) at time t, n_t/N, and the parameter β, which is dependent upon factors such as the infectiousness of the disease (attractiveness of the innovation) and the frequency of contact, both of which are assumed to be fixed [105, pp. 9–10].

For constant β the number of adopters at any time t is clearly a function of the number that have already adopted the innovation, so that a basic characteristic of the process is imitative behavior, or a bandwagon effect [459, p. 77]. However, the absolute increase in adopters at any point in time is the product of opposing forces, since as the proportion that has already adopted, n_t/N, increases, the number of potential adopters, $N - n_t$, falls. This suggests a bell-shaped frequency distribution for numbers adopting over time.

The solution to Eq. (1) is:

$$n_t = N\{1 + \exp(-\alpha - \beta_t\}^{-1} \tag{18}$$

where α is the constant of integration. This is the cumulative density function of the logistic frequency distribution [522, pp.

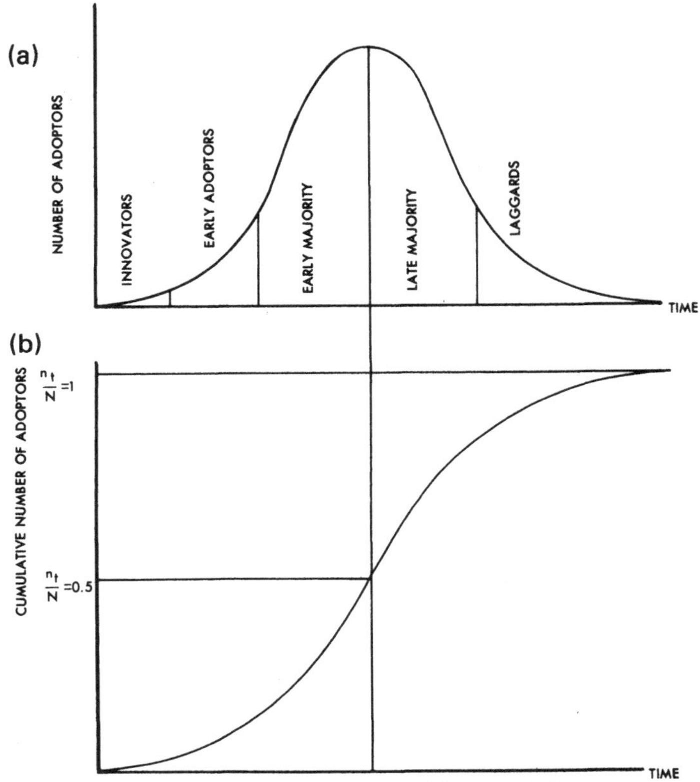

FIGURE 3.1 (a) Adoptor categorization (b) The logistic curve.

69–70] shown in Figure 3.1a and is the equation of the sigmoid (S-shaped) logistic curve shown in Figure 3.1b.[94]

The curve is described by three variables. N is the upper limit, the ceiling approached when the process is completed; β may reasonably be called the speed of diffusion, though it is *not* the rate of growth[95] of diffusion [105, p. 11]; and α, the constant of

[94] Yeomans [579, Ch. 5] provides a general discussion of mathematical trend curves that explains the logistic in a forecasting context. Davies [106, p. 158] points out that though the tails of the two curves are slightly different, it is reasonable to assume a "rough equivalence" between the logistic and the cumulative normal curve.

[95] See also the derivation and discussion of the logistic curve in van Duijn [550].

integration, positions the curve on the time axis. The curve is symmetric around the inflection point, which occurs at time $-(\alpha/\beta)$ corresponding to 50% adoption, and approaches zero and N asymptotically, as t tends to minus and plus infinity.

Though several methods of fitting the logistic curve have been investigated [107, Ch. 11; 379, 550], most empirical investigations are straightforward, using linear regression analysis on the transformation of Eq. (18).

$$\log\left(\frac{n_t}{N - n_t}\right) = \alpha + \beta t \tag{19}$$

This approach forms the basis of the first stage of Griliches' 1957 [190] study of hybrid corn and Mansfield's [313] investigation of twelve innovations in American industry.

3.2. Applications of the epidemic model

So far, the term *diffusion* has been considered only by analogy to the spread of infectious diseases. However, if the object of interest is a process innovation (often the spread of a new capital good), a reasonable measure of the extent of its diffusion would be either the proportion of the post-diffusion capital stock (S_t) currently accounted for by the new machines (s_t) or the proportion of the industry's output currently produced with the new process. Following Mansfield [313], Davies [105, p. 6] calls this concept the *overall* rate of diffusion.

The overall rate depends on both the rate of imitation (inter-firm diffusion), i.e., the proportion of firms that have adopted the innovation, *and* the level of use within each firm, determined by the intra-firm rate of diffusion—the rate at which particular firms substitute the new technology for the old once they have begun to use it [316, p. 173]. Following this convention, we consider first the overall rate of diffusion. Later, economy-wide and international diffusion are considered.[96]

3.2.1. The overall rate of diffusion. One of the best-known

[96] A concise account of applications of the logistic curve to product and process innovations is provided by Bain [23, Ch. 2]. Davies [105] and Stoneman [522] arrange their surveys according to the level at which diffusion is being studied. This approach is avoided here since it leads to repetition.

diffusion studies in economics is Griliches' 1957 [190] investigation of the percentage of U.S. corn acreage planted with hybrid seed. The diffusion of agricultural technology had been intensively studied previously by rural sociologists (summarized in Rogers [423, pp. 57–59] and Summers [527]). Indeed, a particularly influential paper by Ryan and Gross [458] found that the diffusion curve for hybrid corn in Iowa followed a sigmoid pattern.

Griliches discovered graphically the same S-shaped pattern for individual states. He found the trends to be so strong that individual observations could not be explained by economic variables, as if they "had no antecedents." His solution was to fit logistic curves, not because any underlying model justified the logistic, but because curve-fitting reduced the mass of data for each state to just three parameters: origin, rate, and ceiling. In a second-stage analysis, these parameters were explained in terms of the profitability of adoption.

The logistic curve was chosen because it was the easiest to fit and interpret. Thus Griliches used ordinary least squares to fit the transformation of Eq. (18), shown as Eq. (19) above, to time series data for each of 31 states. The values of N_i ($i = 1, \ldots, 31$), the satiation levels for the 31 states, were chosen by visual inspection to give the best fit. Both the upper and lower tails of the distribution were excluded. At the lower extreme this was done by taking the point of origin to be 10% of the ceiling acreage.

This procedure generated estimates of β_i (the rate of diffusion), α_i (the origin, or year at which 10% hybrids were planted), and N_i (the final ceiling level), which were then "explained" by economic variables. The date of origin, α_i, was taken to represent the *supply side* of the problem, with the lag (relative to Iowa) before suitable hybrids became available being explained by varying profitability to seed producers (profitability diminished with distance from the Corn Belt). Differences in the ceiling level of use and speed of diffusion were attributed to demand factors and explained by the profitability of the shift from open pollinating seed to hybrid varieties.[97] Profitability was assumed to depend on corn acres per

[97] Bogue [65, p. 24] raises the point that by the logic of Griliches' model, the fundamental breakthroughs in the development of hybrid corn should have been made at Corn Belt experiment stations, not in Connecticut, an area of "low market density." However, the key individuals had previously worked on corn breeding in the Midwest.

farm, pre-hybrid yield, and the difference in yields between open pollinating and hybrid varieties.

Although Griliches succeeded in explaining a large proportion of the variance in the three parameters, a controversy resulted from his assertion that the variables considered by sociologists "tend to cancel themselves out, leaving the economic variables as the major determinants of the pattern of technological change" [190, p. 522]. In reply, Brander and Straus [68] produced evidence that in the case of hybrid sorghum adoption in Kansas, compatibility[98] appeared to be more important than profitability, and Havens and Rogers [212] and Rogers [420] argued against the importance of profitability in the case of hybrid corn in Iowa. Babcock [21] suggested that the economic and sociological explanations are complementary, and Griliches himself [192, 193, 195] responded by suggesting that the alternative explanations represented a "false dichotomy."[99]

3.2.2. The inter-firm rate of diffusion. Most diffusion studies consider the spread of an innovation among the firms in an industry (inter-firm diffusion) separately from the level of use within the firm (intra-firm diffusion). Mansfield's [313] seminal paper, which analyzed the inter-firm diffusion of twelve innovations in four U.S. industries, constitutes the conventional wisdom in the field [105].

Following Griliches, the diffusion process is treated as an initial situation of disequilibrium created by the innovation, which is corrected by the spread of the new technique up to a new equilibrium level of satiation. The model is developed from the initial proposition that the proportion of *non-users* who adopt the innovation in a given time period will increase with the profitability

[98] Rogers [423, p. 211] lists the attributes of innovations as (1) relative advantage (which includes profitability), (2) compatibility, which is consistency with "existing values, past experience and needs of adopters," (3) complexity, (4) trialability, and (5) observability. The basic argument is that less complex, observable innovations that can be tested first will be more readily adopted than those that lack these attributes.

[99] Griliches also noted that the disagreement is in part semantic. Terms such as *compatibility* or *congruence* can be translated into economic variables with the inclusion of imperfect information and risk preference, which are not at odds with profitability. In the original paper [190, p. 522], Griliches argued that area characteristics and personal characteristics are highly related. That is, low yields are often correlated with low education, low status, low income, and other socio-economic variables.

of the innovation (π) and the proportion of firms that have already adopted (n_t/N), but will be inversely related to the size of the investment outlay required (S), giving the equation:

$$(n_{t+1} - n_t)/(N - n_t) = f\left(\frac{n_t}{N}, \pi, S\right). \tag{20}$$

The function f is approximated by a Taylor's series expansion with third and higher order terms ignored, along with a quadratic term in (n_t/N). If the time period is sufficiently short, Eq. (20) may be written in differential equation form as,

$$\frac{dn_t}{dt}\left(\frac{1}{N - n_t}\right) = \beta_0 + \beta_1 \frac{n_t}{N} \tag{21}$$

where

$$\beta_1 \frac{n_t}{N} = a_0 + a_1\pi + a_2 S + \varepsilon \tag{22}$$

and ε represents the error structure.

Assuming that the limiting value of n_t is zero as t approaches negative infinity, Eq. (21) has the solution,

$$n_t = \frac{N}{1 + e^{(-\alpha-\beta t)}}, \tag{23}$$

which is the logistic curve.

This result is hardly surprising, since once the limit condition has constrained β_0 to equal zero, the differential Eq. (21) is clearly the equation of the epidemic model (see Eq. (17) above). This has led Davies [105, p. 15] to argue that the model is no more than an ingenious application of the epidemic model and that no economic content would be lost in taking Eq. (21) as the starting point, since it is obtained from Eq. (20) by assumption and algebraic manipulation.

Davies [105, pp. 17–18] is also critical of Mansfield's fitting of Eq. (19) above by weighted least squares. The coefficient of correlation between the dependent variable and time exceeded 0.89 in all cases, suggesting that the logistic curve does fit the data well. However, the estimates are based on an average of only ten observations per innovation, and Mansfield's exclusion of smaller firms from the sample because of lack of information may lead to bias. In their study of 13 innovations in the United States, Gold *et al.* [177] find such a diversity of variables and "special circumstances" that they are critical of "universal" models like Mansfield's. They suggest a broader analytical framework and concentrate on investigating the firm's decision-making process.

The second stage of Mansfield's study attempts to explain the estimates of β_1 for individual innovations by cross-section estimation of Eq. (22). Mansfield found that the coefficients (a_1 and a_2) of both independent variables were significantly different from zero and had the predicted signs. Though the equation fitted extremely well, the sample size of 12 innovations is too small for comfort. Further results suggested that diffusion was faster, the less durable the industry's capital equipment and the greater the growth rate of output.

Mansfield *et al.* [318] and Mansfield [317] have also applied the epidemic model to numerically controlled machine tools. Globerman [173] fitted the same model to Canadian data, allowing comparisons with both the Mansfield results and the further U.S. evidence generated by Romeo [426, 427]. Romeo found interindustry differences in the speed of diffusion of numerically controlled machine tools to be partially explained by the number of firms in the using industry (positively) and the variance of the logarithm of firm size (negatively), suggesting that diffusion speed is increased by competition. Stoneman [522, p. 95] is critical of the small sample size and Romeo's version of Eq. (22), which is written in multiplicative form without explanation.

Although the authors make little attempt to extend the theory, the collection of papers in Nasbeth and Ray [350] provides a wealth of evidence on the diffusion of industrial processes in Western European nations. Particularly relevant are the papers by Lacci, Davies, and Smith applying the logistic curve (and alternatives) to tunnel kilns in brick-making and gibberellic acid in malting. Ray [410] has provided more results, updating the original studies.

Within the large marketing literature on technological forecasting and product innovations (see Section 3.32), Mansfield's model has been applied and developed by Blackman [58], Fisher and Pry [159], and Sharif and Kabir [492]. More recently, the Mansfield–Blackman model has been extended by Mahajan and Peterson [309] to integrate diffusion over both time and space (i.e., market regions). Sharif and Haq [491] identify no less than nine explanatory variables that can be added to Mansfield's list of factors explaining the rate of diffusion. Ayres' [20] "Schumperterian" model of diffusion and profitability provides an alternative to Mansfield's approach.

3.2.3. Intra-firm diffusion of technology. On intra-firm diffusion the standard reference is also work by Mansfield [314, 316, Ch. 9], which applies the methodology described above to the spread of diesel locomotives within thirty U.S. railroad companies between 1925 and 1960. The paper begins with the theoretical proposition that the increase in firm *i*'s stock of diesels at time *t* as a proportion of the additions still to be made,

$$\left(\frac{s_{i(t+1)} - s_{it}}{S_i - s_{it}}\right),$$

will vary positively with expected profitability (π_i) and the firm's liquidity (C_i), negatively with the apparent risk (U_i), and also with firm size (I_i) (in a manner not specified *a priori*). Then, assuming risk to be lower for late adopters and to decline as satiation (S_i) is approached, U_i can be represented by the date of adoption (L_i) and the proportionate level of adoption,

$$\left(\frac{s_{it}}{S_i}\right),$$

Written in differential equation form, this gives:

$$\frac{ds_{it}}{d_t} = g(\cdot)(S_i - s_{it}), \tag{24}$$

where $g(\cdot)$ is a function of the variables listed above:

$$g(\cdot) = \phi_i \frac{s_{it}}{S_i} \tag{25}$$

where

$$\phi_i = a_i + a_2\pi_i + a_3L_i + a_4I_i + a_5C_i + \varepsilon_i. \tag{26}$$

Manipulations fully described by Stoneman [572, p. 75] lead directly to the logistic curve equation:

$$s_{it} = \frac{S_i}{1 + e^{-(\phi_i + \alpha)}}. \tag{27}$$

Fitting the linear transformation of the logistic curve (using the weighted least squares method of Berkson [38]) generates estimates (ϕ_i) of the speed of diffusion for each firm. As in Mansfield's inter-firm study, described in the last subsection, these estimates become the dependent variables in cross-section estimation of Eq. (26) above. The estimated coefficients (a_i) had the predicted sign and were significant for all variables except I. About 70% of the inter-firm variation in the rate of dieselization is explained by this second stage analysis.

In criticizing Mansfield's empirical methods (as opposed to theory), Stoneman [522, pp. 74–85] questions the manner in which the error term ε_i (see Eq. (26)) should be included in a logistic model. He argues that Berkson's weights and methodology may not be applicable to Mansfield's model.

Much of Stoneman's critique extends to the work of Romeo [426], who followed Mansfield's approach in studying the intra-firm diffusion of numerically controlled machine tools. Romeo's model differs from Mansfield's in taking as the dependent variable the proportion of the firm's *new machine tool purchases* at time t that are numerically controlled. Also, in the second stage analysis, the variables that explain the rates of diffusion are assumed to be linear in logarithms, rather than simply linear. Further empirical evidence on machine tools, some of which supports Romeo's assumptions, is provided by Nasbeth and Ray [350]. They also analyze the intra-firm diffusion of special presses, the basic oxygen process, and continuous casting in steel. Although in many of these studies, curve-fitting is not actually attempted and the theory is not developed beyond the epidemic model [105, p. 26], they do provide detailed evidence for the diffusion of ten major process innovations in six countries, which facilitates international comparisons.

3.2.4. International diffusion. Mansfield's methodology has been applied to the diffusion of synthetic rubber in 12 countries by Swann [529], who explains the parameters of fitted logistic curves for each nation by means of country-level variables such as output growth, rubber imports, rubber exports, and production of rubber per capita.

Neither Swann nor Nasbeth and Ray really tackle the problem of the *transmission of technology between countries*. Some aspects of international technology transfer are briefly considered in Section 3.7.

3.2.5. Nonprofit firms, regulated industries, and the public sector. The U.S. hospital sector[100] includes a considerable proportion of non-profit firms, whose motivation for adopting new techniques must differ from the profit-orientation described above. Instead, improvement in the quality of medical care may be the primary motivation. However, Rapoport's [409] study of the diffusion of radioactive isotope use in U.S. hospitals compares parameters for the speed of diffusion across different states (using the results from fitted logistic curves) and concludes that the speed is greater where the environment is competitive, a result no different from the conventional wisdom for profit-seeking firms [263, 477].

The effect of regulation on the diffusion of innovations has been investigated by Capron [79]. Oster and Quigley [385] considered the effect of building code regulations on four innovations in house-building, finding that labor-saving innovations were less rapidly adopted than others. Key variables affecting diffusion speeds among jurisdictions were the extent of unionization, firm size, and the professionalism of local regulators (measured by education, background, and professional contacts). However, regulation should not be assumed automatically to retard diffusion. Sweeney [530] demonstrates the well-known result that in a situation of cost-plus-markup regulation, the existence of regulatory lag (the interval between cost reduction and price reduction) provides an incentive to the firm to adopt cost-reducing innovations to obtain excess profits. Ironically, inefficiencies in implementing the regulation system can encourage technical efficiency. Lastly, an appraisal of innovation and diffusion in the public sector is provided by Feller and Monzel [146].

3.3. Alternatives to the epidemic model

Though in several cases the logistic curve has provided a useful means of quantifying the diffusion process, it is merely one of a

[100] The diffusion of hospital technologies is considered by Russell and Burke [442] and Russell [441].

large class of S-shaped curves. Indeed, any unimodal frequency distribution will have a sigmoid cumulative density function, which need not be symmetric. Frequently, observed diffusion patterns exhibit an element of skewness and may be better represented by asymmetric functions such as the Gompertz curve or the cumulative lognormal.[101]

3.3.1. Asymmetric diffusion curves. The Gompertz is positively skewed with the inflexion point at $n_t/N = 0.37$. The cumulative lognormal can reproduce a whole family of S-shaped curves, since the inflection point is variable. It corresponds to the frequency distribution's mode, which is defined by $\mu - \sigma^2$, thus depending on both the mean and the variance of the distribution.[102]

However, no one, general form can represent all sigmoid curves as special cases. Nor can one general form subsume the most common candidates such as the logistic and the Gompertz [522, p. 71].[103] One curve may fit better than another or several may appear to fit equally well, making it impossible to discriminate on empirical grounds [459, p. 78]. Griliches [190] did not "want to argue the relative merits of the various S-shapes" and treated the curve-fitting stage of his analysis purely as a means of concentrating the data. Most later writers argue that the form of the curve should be determined by the theory of the diffusion process.[104] Thus, the

---

[101] The Gompertz curve may be expressed as:

$$\frac{dn_t}{dt} = \beta_1 n_t(\log N - \log n_t), \tag{A.1}$$

which is similar to the logistic (see Eq. (1)). If $y = \log x$ is distributed normally, $x = e^y$ has the lognormal distribution. Maddala argues that "since many variables in economics cannot take negative values, and also do not have symmetric distributions as the normal, the lognormal distribution may be more appropriate in some economic applications than the normal" [304, p. 33]. The linear approximation of the lognormal is given by,

$$\frac{dn_t}{dt}\frac{1}{n_t} = \frac{\beta_2}{t}\left(\frac{N - n_t}{N}\right), \quad \beta_2 > 0 \tag{A.2}$$

[526, p. 25].

[102] Cramer [95, pp. 30–31] comments that use of the normal distribution can always be defended by appeal to the central limit theorem.

[103] A survey of studies using asymmetric curves can be found in Bain [23, Chs. 2 and 3], who fits a lognormal curve (also used by Davies [105]) to data for television ownership in the U.K. Dixon [118], discussed below, applies the Gompertz (see also Hernes [224]) to hybrid corn data.

[104] While this view may appeal to economists, some of the diffusion functions used in marketing literature deliberately impose no prior constraints on the data.

appropriate form can be chosen on a priori grounds, and the curve-fitting exercise serves as an empirical test of the particular diffusion hypothesis.

The simple epidemic model described above rests on stringent assumptions concerning the population, the innovation, and the method of transmission. By relaxing these assumptions[105] and by tackling economic issues from which the epidemic approach diverts attention, it is possible to generate a wide range of diffusion models, several of which justify fitting non-symmetric functions.

To an extent, the epidemic model has been infused with economic content but remains unsatisfactory since only the demand side of the problem is included. The only economic issue that is taken up is the possible profitability of the innovation to potential adopters. Casual empiricism suggests that many innovations are supplied by firms that go to considerable trouble to ensure consumer awareness and availability of their products. Similarly, public agencies may induce innovation diffusion as a matter of policy. For example, the USDA's Federal Extension Service is the world's largest public investment in diffusion [423]. Brown's [70] "market and infrastructure perspective" is introduced with a solid argument for recognizing the supply side. He reasons that first, diffusion agencies must be established, and second, a diffusion strategy must be implemented. These two elements of the process precede the actual adoption of the innovation. The next section shows how inclusion of supply side factors can explain positive skewness of the diffusion curve.

3.3.2. Incorporating diffusion from a constant source: the generalized static model. The assumption that diffusion depends on demonstration effects and learning from the experience of others is crucial to the epidemic/logistic model. But personal interaction between adopters and potential adopters may be unimportant. Instead, the innovation may be diffused from a "constant source." That is, the firm selling the innovation may rely on mass media propagation, salesmen, extension agents, etc. In such a situation,

[105] Davies [105, pp. 11–13] demonstrates that if β varies over time, positively and negatively skewed diffusion curves are generated according to whether $\beta(t)$ decreases or increases.

the instantaneous rate of diffusion will decline continually as the gap between the actual and the desired stock decreases at a constant proportional rate [23, p. 9]. Diagrammatically, the function, variously known as the waning exponential [291] or modified exponential [292], could be represented by the curve to the right of the inflection point in Figure 3.1b, since all that matters is the remaining distance to the saturation level.[106] This exponential curve has "received substantial empirical support" in marketing, where it has been referred to as the Coleman model, and in sociology [292, p. 364]. It has been extensively used in economics to study the demand for durable goods and product innovation (see Pyatt [402] and Bain [23] for reviews of early work on the diffusion of product innovations).

Lekvall and Wahlbin [292] refer to the passing of information by social interaction as the *internal* influence and that conveyed by the mass media or other promotional activity as the *external* influence, emanating from a source outside the group of prospective adopters.[107] They argue that for most innovations, both forces will be present in some combination and that the result will be a positively skewed curve, with the skew being greater for heavily advertized consumer products than for production innovations like a new seed, which would be much discussed among farmers.[108] The "two step flow of communication hypothesis" of Lazarsfeld, Berelson, and Gaudet [288] combines the idea of internal and external sources with a heterogeneous population. In their model a mass media message does not reach most receivers directly, but is first taken up by opinion leaders, who pass the word to others. The effects of heterogeneity of the adopter population were considered

[106] The equation for this "waning exponential" curve is:

$$\frac{dn_t}{dt} = \frac{\beta}{N}(N - n_t). \tag{A.3}$$

[107] Alternatively, in the literature on the economics of advertizing, Gould [183] refers to the exponential case, arising from an impersonal advertizing medium [539], as a *diffusion* model, whereas the epidemic word-of-mouth case discussed by Ozga [386] is called a *contagion* model Gould [183] investigates optional advertizing policies for both types of process.

[108] Bain's [23] survey does show that, whereas early studies of process innovations relied on the logistic curve, several pioneering papers on product innovation did apply skewed curves. Advertizing is much more intensive in consumer goods industries than in the capital goods sector.

in detail by Coleman [92, Ch. 17] and formalized by Davies [105, pp. 12–13], who provides a simple demonstration that a heterogeneous population alone is sufficient to give rise to a skewed diffusion curve. Dividing the population into two groups, with different probabilities of adoption, the differential equation becomes,

$$\frac{dn_t}{dt} = \beta_1\left(\frac{n_t}{N}\right)(N_1 - n_{1t}) + \beta_2\left(\frac{n_t}{N}\right)(N_2 - n_{2t}), \tag{28}$$

which is shown to have an inflection point at $n_t/N < 0.5$ (i.e. a positive skew).

Both external and internal sources and a heterogeneous population are effectively combined in the "influential *new product*[109] growth model" for consumer durables introduced by Bass [28], and developed by Mahajan and Schoeman [312] and others. If the rate of diffusion is taken to be proportional to the number of potential adopters available, then a general form for the differential equation is

$$\frac{dn_t}{dt} = g(t)(N - n_t). \tag{29}$$

If $g(t)$ is a simple linear function of the number who have adopted to date, then

$$g(t) = \beta_0 + \beta_1 n_t. \tag{30}$$

Substitution gives,

$$\frac{dn_t}{dt} = (\beta_0 + \beta_1 n_t)(N - n_t), \tag{31}$$

which may be written as

$$\frac{dn_t}{dt} = \beta_0(N - n_t) + \beta_1 n_t(N - n_t), \tag{32}$$

which combines Eqs. (17) (the logistic) and the waning exponential[110] (Eq. A.3 in footnote 106).

The constant β_0 is the coefficient for the proportion of the

[109] Stoneman [522] argues that the general principles of analysis applied to process and product innovations are very similar.

[110] Skiadas [499] describes an equation of this form as a linear combination of the Blackman/Fisher–Pry and Coleman models. His paper provides a useful list of equations for many of the models discussed here, supported by graphs of their frequency distributions.

population whose adoption decision depends on information from a central source. Similarly, β_1 is the coefficient of imitation, since the second term reflects adoption due to personal interaction [310, p. 130].

Equation (32) reduces to the logistic when $\beta_0 = 0$ and to the exponential when $\beta_1 = 0$. In all intermediate cases, it will produce an S-shaped curve that will mirror the skew of the data. In their useful survey of the development of new product diffusion models, Sharif and Ramanathan [493] refer to this approach as the "generalized model" while attributing the logistic approach to Dodd [119] and the decaying exponential representation to Coleman [92].

Since Eq. (32) can be written as

$$\frac{dn_t}{dt} = a_t = \beta_0 N + (\beta_1 N - \beta_0)n_t - \beta_1(n_t)^2, \tag{33}$$

Bass suggests the estimating equation,

$$a_t = \gamma_0 + \gamma_1 n_t + \gamma_2(n_t)^2 \tag{34}$$

where

$$\gamma_0 = \beta_0 N, \quad \gamma_1 = \beta_1 N - \beta_0 \quad \text{and} \quad \gamma_2 = -\beta_1.$$

Bass tested the model, using time series data for 11 consumer durables. He found that the model generally performed well in forecasting the magnitude and timing of the peak level of sales.

The manner in which the "generalized model" shown in Eq. (32) has been developed is summarized by Mahajan and Peterson [310]. Static models retain the fixed number of potential adopters (N) and operate on the function $g(t)$ in Eq. (29), whereas dynamic models make N a function of relevant time-dependent variables (see next Section). Thus Robinson and Lakhani [419] follow the first route, arguing that to enable firms to evaluate marketing strategies, β_1 must be developed as a function of the decision variables, such as price, advertizing, and promotions. Similarly, Horsky and Simon [233] model β_0 in Eq. (32) as a function of advertizing expenditures, and Bass [29] makes both coefficients functions of the level of adoption, n_t. Numerous permutations are defined by Mahajan and Peterson [310].[111] The main point is that the form of β_0 and β_1 in

[111] A complete survey of the development of diffusion models in the marketing literature, including several applications, is now available in Mahajan and Peterson [311].

Eq. (32) implicitly assumes that the company or diffusion agency does not change its behavior during the diffusion period or product life cycle. Another solution to this problem is provided by Lilien [295], who incorporated a control variable by which the agency can influence the diffusion process.

Several other developments have helped to make diffusion models more realistic. For example, Sahal [459, p. 81] has stressed that "the diffusion of an innovation does not take place in isolation. Rather, it is very much a matter of actual substitution of a new technique for the old." Thus, Sahal expands a relatively early development of Mansfield's model by Fisher and Pry [159] in which the rate of adoption of a new product is proportional to the level of use of the old product being replaced. That innovations do not exist in isolation has also been addressed by Mahajan and Peterson [308], who include equations for the new product and an existing good that may be independent, complementary, contingent, or substitutable for the old good. Sharif and Kabir [492] apply the techniques of system dynamics in developing a "multilevel" model of technological substitution in which a particular product or technology is replacing an older one while it is itself being replaced by a still newer product or process.

The models considered above can be called "binomial" in the sense that the population is divided into two groups, adopters and potential adopters. Such models implicitly assume that the entire population eventually adopts the innovation and that, once adopted, the innovation is never rejected. These assumptions are avoided by Sharif and Ramanathan [494], whose "polynomial" model divides the population into four groups. These are adopters, rejectors, disapprovers, and the remainder, who are as yet uncommitted. The approach is also "multilevel" (as in Sharif and Kabir [492], above), analyzing the substitution between an old, an intermediate, and a new product. This allows the full product life cycle to be modeled explicitly, with the empirical example showing the decline of black and white television as well as the growth of ownership of colour sets.

In Griliches [190] the spatial, as well as the temporal, aspects of the diffusion process are quite explicit. The work of the geographers, reviewed by Brown [70], has tended to build on the studies by Hagerstrand [205], refining the Monte Carlo simulation models he pioneered.[112] Both the temporal and spatial models of the phenomenon are considered by Sahal [459, Ch. 5], who discusses two "complementary" models. The temporal model appears best to represent technological substitution in cases where the adoption process is measured in terms of annual sales of the new product, whereas the spatial model is more appropriate where the diffusion data refer to a stock variable.[113] Mahajan and Peterson [309] integrate the time and space dimensions of the diffusion process in a model that incorporates the distance of the subject region from the location where the product is first introduced. This "neighborhood effect," whereby the "innovation waves" spread from the center to the periphery, is added to the general model.

To summarize, the "general static diffusion model" incorporates diffusion both by word of mouth and diffusion from a central source. It has been expanded to include explicitly the effect of economic variables such as product prices, advertising expenditures, and demonstration efforts.[114] The technological substitution process has also been explicitly incorporated, including the case where several technologies are involved. The population has been divided into more than two groups, allowing for rejection of the innovation and the effect on the diffusion process of persons who actively disapprove of the innovation. Lastly, geographical space has been included in the models. We now turn to the work of authors who have rejected the assumption of a fixed population of adopters.

3.3.3. Dynamic models. While the "generalized static model" (Eq. (28)) can mirror any degree of skewness in the data and provide reasonable estimates for technological forecasting, it may

---

[112] See also Brown and Lentnek [71] for an example of a spatial diffusion study with economic content. In spatial diffusion, Grigg [189, Ch. 11] draws a distinction between "migration" diffusion and "stimulus" diffusion.

[113] In the marketing literature particularly, the reader should be wary of the distinctions between *stocks* and *flows* such as sales, between durable and non-durable goods, between first-purchase and replacement sales. See particularly Olsen and Choi [381], whose graphs decompose total sales into first-purchases and repeat buying, showing the curves to be very different.

[114] These models include aspects of supply in the diffusion equation, often by making a coefficient a function of the (exogenous) supply price or advertizing expenditure. The distinction can be a little arbitrary, but really they do not attempt to model the pricing and output decisions of the suppliers of new technology. Studies that do have an explicit side are considered in Section 3.6.

be misleading even in this application. The most obviously unrealistic assumption is that of a fixed population of adopters and, by implication, no post-innovation improvements. This is unfortunate, since the historical literature reviewed by Rosenberg [435, Ch. 1] suggests a "view of technical progress as consisting of a steady accretion of innumerable minor improvements and modifications, with only very infrequent major innovations" (p. 7).[115] This emphasis on the importance of follow-up improvements is supported by studies such as Enos [128] on petroleum technology, Miller and Sawers [338] on aircraft, and Sahal [459] on tractors and other machinery.

Kuznets [283, pp. 337–338] divides the product cycle into phases called "initial application," "diffusion," and "slowdown and obsolescence." He argues that for many innovations, such as television, automobiles, and computers, substantial ongoing quality and cost improvements cause extreme difficulty in defining the end of the diffusion phase in terms of an upper limit in numbers.

The problem is well illustrated in a recent paper by Dixon [117] repeating Griliches' [190] work on hybrid corn and especially by Griliches' [195] reply. Dixon found that in 21 of the 31 states studied, a positively skewed diffusion curve (the Gompertz) was "more apt" than the logistic curve used by Griliches. However, the skewness arises largely because data subsequently available showed that adoption had exceeded the "ceilings" assumed by Griliches. Griliches [195, p. 1463] replied that he would not fit an asymmetric curve but "would now respecify the model so that the ceiling is itself a *function of economic variables* that change over time." This proposition is developed further by Metcalfe [332, pp. 349–350], who argues that "instead of a single diffusion curve, we have an envelope of successive diffusion curves, each appropriate to a given set of innovation and adoption environmental characteristics, each with its own value of N and β. While any given set of characteristics generates a logistic process, the envelope need not conform to the logistic pattern and its exact shape will depend on the temporal

[115] Rosenberg [432, Ch. 11] discusses improvements to innovations and other factors affecting diffusion that are difficult to model. One intractable difficulty is that innovation is continuous and the potential adopters' technological expectations will crucially affect the adoption decision [436, Ch. 5]. Technological expectations have been modeled by Balcer and Lippman [24].

incidence of the changes in its characteristics." A model of this type for consumer durables has been developed by Mahajan and Peterson [308]. The endogenous and shifting ceiling is incorporated by including in the equations the growth of housing starts.[116] The result is a "product growth curve" that is the envelope of the appropriate portions of a whole series of diffusion curves for a series of increasing ceilings. Over time, this product growth curve approaches the cumulative market potential curve and finally coincides with it.

Hernes [224] criticized Lekvall and Wahlbin for attributing the skewness of the diffusion curve to "external and internal forces" and offered the following alternative classification of causes: first, structural heterogeneity, when some variable such as purchasing power is differentially distributed among the population; second, dynamic heterogeneity, when the population changes during the diffusion process (rising income levels would be an obvious example); and last, changing stimulus over time, which includes changes in the quality of the product itself. His own model incorporated the last of these three factors.[117]

By now it should be apparent that there are fundamental difficulties with the basic approach to diffusion followed thus far. It assumes an economic system in which the original equilibrium has been disturbed by the introduction of an innovation. The diffusion process is viewed as the adjustment from the old to the new equilibria. Adjustment is not instantaneous because of the asymmetric distribution of information. Griliches [195] explains that "if all variables describing individuals and affecting them were observable, one might do without the notion of diffusion and discuss everything within an equilibrium framework. Since much of the interesting data are unobservable, time is brought in to proxy for at least three sets of distinct forces." There are (1) declining real costs of the technology due to cumulative improvements and learning by doing (or by using); (2) the scrapping of old durable equipment,

---

[116] Just as car ownership influences tire production, increasing the housing stock increases the stock of new consumer durables. United States housing starts had tripled between 1950 and the time of writing, making the static model inappropriate [308].

[117] A further contribution by Dodson and Muller [120] explicitly introduces advertizing expenditures in a dynamic model.

making way for the new; and (3) risk reduction due to the spread of information about the operating characteristics, workability, and profitability of the new technology. As Griliches suggests, the alternative to his own disequilibrium approach is to model these economic determinants explicitly.[118] We turn now to that challenge beginning with vintage capital (Griliches' second point) and stock adjustment models.

3.3.4. Vintage and stock adjustment models. The vintage capital model explains the co-existence of old and new techniques, allowing the time element to enter the model while retaining the notion of equilibrium. Indeed, Salter's [462] book, which is the example discussed by both Davies [105] and Stoneman [522], calls the vintage approach "a model of the delay in the utilization of new techniques of production."[119] The "delay" is caused because only the plants most recently built will embody the latest technology appropriate for current factor price ratios. With a continuous stream of technical advances, the spectrum of plants in existence at any one time provides a fossilized history of technology. New "best practive" plants[120] will be added to the leading edge of this spectrum and plants at the trailing edge will be scrapped, but only when revenues no longer cover variable operating costs.

Despite its obvious attractions, the vintage model has not been fully developed as a framework for the study of diffusion.[121] However, it has been applied by Sumrall [528] in conjunction with the ideas of Tobin and Brainard [544] on investment theory. Sumrall's model adds to earlier studies, comparing rates of return on five vintages of basic oxygen furnaces (BOF), electric furnaces,

[118] Stoneman [522, Ch. 5] provides a detailed, if abstract, comparison of the views of diffusion as an equilibrium or disequilibrium concept.

[119] This is actually the title of Chapter 4 of Salter's Book [462], in which the model is developed. See also Nasbeth and Ray's verbal account of the economic reasons for a slow start to the diffusion process, neatly summarized in Kelly and Kransberg [267, p. 125]. Brown [70, pp. 191–192] offers an alternative set of reasons for a slow start.

[120] That new investment is assumed to be in machinery of the latest type is a disadvantage, since the evidence suggests that this is often not the case. See especially Gregory and James [187] and Gomulka [179].

[121] Davies [105, p. 30] argues that threshold models such as the one developed by David [101] rest on a view of decision making that is not dissimilar from Salter's vintage model. Models of this type are considered in Section 3.5. See also David [102], who frequently refers to Salter's book.

and open hearth furnaces, testing the hypothesis that "firms adopted the BOF at an optimal rate." The hypothesis is rejected, leading to the conclusion that large firms trailed their smaller counterparts in adopting the BOF. This finding is incompatible with the Schumpeterian hypothesis that large firms are more technologically progressive.

The investment theory concept most extensively applied to the diffusion of product innovations is the stock adjustment model (see Stone and Rowe [518] for an early example). More recently, the model has been applied to the diffusion of computers by Chow [89] for the United States and by Stoneman [519] for the United Kingdom. Beginning from the epidemic approach, Chow [89, p. 1118] postulates that the growth of computer usage at time t will be proportional to the difference between the actual stock n_t and the equilibrium stock N_t. Chow suggests two alternative differential equations,

$$\frac{dn_t}{dt} = \beta n_t(N_t - n_t) \qquad (35)$$

and

$$\frac{dn_t}{dt} = \beta n_t(\log N_t - \log n_t), \qquad (36)$$

which yield as their solutions the familiar logistic and Gompertz curves (see Section 3.1 and footnote 101). Though the similarity to the epidemic model is clear, the stock adjustment does give the model a theoretical base. Stoneman [522, pp. 115–117] shows that diffusion curves can easily be derived using the stock adjustment concept. He provides a simple derivation of the logistic equation by assuming maximization of profits subject to a given production function constrained by adjustment costs. Both models are improved by dynamic formulations. N_t is not a constant, but a function of the relative price of the new technology and the level of GNP. In addition, Stoneman defines β (the speed of adjustment) as a function of economic variables. The better results of both the U.K. and U.S. studies, derived by fitting Gompertz curves, are reported and discussed in Stoneman [522, pp. 135–140].

In terms of the classification system developed above, both studies are of the "overall" rate of diffusion at the economy-wide level, making no distinction between inter-firm and intra-firm diffusion, which Davies [105] regards as a serious limitation. However, models based on investment theory clearly do provide an economic explanation of the delay involved in the diffusion of techniques. So too can macroeconomic growth theories, as Hicks [228, Ch. 2] has shown. Much new process technology is embodied in investment goods. The rate at which investment can proceed must be limited by the rate at which savings can be increased and by factor shortages, which will themselves give rise to technical changes. These are Hicks' "induced inventions" discussed in Section 2.2 above.

3.4. Adoption studies

In this section we review the literature on differential adoption.

3.4.1. The social–psychological tradition. Since the diffusion curves described in early sections are intended to model the behavior of firms in aggregate, it is inevitable that they cannot explain why some firms adopt innovations faster than others [105, p. 15]. This was recognized by Mansfield [315, 316, Ch. 8], who investigated "the speed of response of individual firms."[122] He argues that the length of time a firm waits before using a new technique tends to be inversely related to its size. A large firm is better able to handle the costs and risks involved and is more likely to have both an early need to replace equipment and operating conditions suited to the new technique. Similarly, the length of time before adoption may be expected to be inversely related to the profitability of adoption by the firm.

Assuming that the relationship is multiplicative, these propositions lead to the testable hypothesis,

$$d_{ij} = Q_i S_{ij}^{a_{i2}} H_{ij}^{a_{i3}} e^{u_{ij}} \tag{37}$$

where d_{ij} is the number of years the jth firm waits before using the ith innovation, S_{ij} is its size, H_{ij} is a measure of the profitability of the investment in the innovation, u_{ij} is a stochastic error term, the

[122] The "second stages" of the studies by Griliches [190] and Mansfield [313, 314] explain the rates of diffusion for different areas and industries and *within* different firms. Here, we consider individual *firms within one industry,* rather than comparing industries.

a_i's are parameters, and Q_i is a scale factor that varies across innovations. Equation (37) was fitted to data on 167 firms for 14 innovations. Measurement difficulties allowed H_{ij} to be included for only five of the innovations, and it was statistically significant for only two of these. However, the coefficient of S_{ij} is consistently negative and statistically significant, though Davies [105, pp. 22–23] shows that interpretation is difficult. For instance, this result would emerge if industry *A*, with a few large firms, adopted an innovation more quickly than industry *B*, with many small firms, even if there were no correlation between early adoption and firm size *within* either industry.

Mansfield extended the empirical analysis to include as explanatory variables the firm's rate of growth, its profitability, the age of its president, its profit trend, and a measure of its liquidity. He also allowed for different types of innovation (for instance, highly costly or relatively cheap). Unfortunately the coefficients were not statistically significant.

Mansfield's approach has been applied to several industries by other authors. Examples are Nasbeth's study of six innovations in Sweden (in Williams [564]); Hastings' [211] investigation of the adoption of four process innovations in the Australian wool textile industry; Oster's [384] study of the adoption of the basic oxygen furnace in the United States; and Benvignati's [34, 35] studies of adoption in the cotton textile industry. The last of these works reaches the conclusion that the diffusion of domestic innovations is more rapid than for foreign advances. A further source of information on adoption is the Nasbeth and Ray [350] volume, especially the papers by Hakonson and Smith, which study innovations in which the diffusion process is incomplete, and hence not all the d_{ij}'s are known. This problem has also been tackled in the diffusion context by Jarvis [241], who has applied techniques similar to those used by Griliches [190], to the diffusion of pasture improvements in Uruguay.

Both Stoneman [522] and Davies [105] point out that Mansfield's adoption model appears to be unconnected or even inconsistent with his diffusion models and indeed bears no relation to ideas of information-gathering and uncertainty, which underlie his earlier approach. This is unfortunate, since aggregation of the d_{ij}'s for each industry should give the inter-firm diffusion curve. Intuitively, the determinants of adoption at any moment in time should be expected to be the independent variables for which time served as a proxy in diffusion curve-fitting. However, the variables chosen by Mansfield do not seem to reflect adequately the determinants of diffusion suggested by either Hernes [224] or Griliches [195] in Section 3.3.3.

Mohr's [340] methodological contribution offers an explanation of this poor correspondence between diffusion and adoption models. Whereas diffusion models are categorized as process theory, adoption models of the Mansfield type are an example of variance theory. (These terms are defined in footnote 93.) Mohr [340, pp. 57–58] considers the effects of imposing a "degenerate" variance theory were a process theory is appropriate, and his quotations (pp. 68–69) demonstrate the confusion that results from unwittingly mixing diffusion and adoption approaches.

Mansfield's pioneering diffusion and adoption studies defined the conventional wisdom on the subject until recently. Stoneman [522] refers to Mansfield's diffusion model as the "psychological approach," and indeed Mansfield [316] does argue that "there exists an important economic analogue to the classic pyschological laws relating reaction time to the intensity of the stimulus." The "psychological approach" is displayed most strongly in the work on adoption described above, in which "attitudinal" variables and variables intended to take account of the *attributes of innovations* are to be found. In this, the adoption studies more closely resemble the work of other social scientists who have emphasized these "non-economic" variables. Kelly and Kransberg [267, pp. 127–129] consider the "Social-psychological Tradition in Diffusion Research," concentrating on the importance of social networks, studies of resistance to change, and the effects of education on adoption. The discussion of the effects of education shows that attempts at classification are somewhat arbitrary, since they concentrate on the work of Nelson and Phelps [355] and Hayami and Ruttan [219], all of whom would be more at home classified under "the Economic Perspective" (the other alternative is "Spatial Diffusion").

Feller's [147] classification of differing approaches to innovation includes an intepretation of the argument between Griliches and the rural sociologists and a discussion of the importance of the "sociological variables" that determine entrepreneurial attitudes.

Rogers [423] provides a comprehensive summary of the non-economic literature.[123] The core of his book comprises chapters on the adoption-decision process (broken down into five stages), the origin and attributes of innovations (see footnote 98), categorization of adopters, information networks, and the role of the change agent.[124] Adopters are categorized as idealized types, according to time of adoption, as shown in Figure 3.1a. Innovators are described as respectable local opinion leaders. The early majority follow with deliberate willingness, while peer pressure is necessary to convince skeptical late adopters. The laggards are traditionalists who cling to the past.

3.4.2. Agricultural adoption studies. Though the extensive adoption literature outside of mainstream economics cannot be described adequately here, a brief summary of common methodologies follows, drawing examples from agricultural economics, and rural sociology where such studies proliferate.[125] This is because "diffusion is the dominant mechanism for the spread of a new technology in sectors where the firms are small compared with the market as a whole and where, for a variety of reasons, they are unable to expand their market share rapidly. Farming is the archetypical example" [354, p. 1050]. This contrasts with industries in which firms do their own R & D and where the expansion of

---

[123] Economists may be confused by his terminology. Though his book is called *The Diffusion of Innovations,* it is predominantly about adoption. Rogers and Eveland [424, p. 283] make his usage of the term *diffusion* quite clear, stating that "correlation analysis of one-shot survey data is overwhelmingly the favorite methodology of diffusion investigators." In this paper, work of this type has been called *adoption,* whereas non-economists use the terms interchangeably.

[124] There is also a chapter on innovation and adoption in organizations. Although economists do not differentiate between the decision-making processes of individuals and of organizations, Rogers suggests some important differences. For instance, the adoption decision and its implementation may be quite separate processes in an organization. To adequately explain adoption by organizations the work of behavioral scientists must be incorporated [424]. The distinction is made more important by the fact that most new product adoption may depend on individual (or family-group) decision-making, whereas process innovations are usually adopted by organizations.

[125] Rogers [420] provides a comprehensive survey of the earlier literature. Fliegel and van Es (in Summers [527]), give a brief history of the adoption and diffusion literature in the particularly active area of rural sociology. Feder, Just, and Zilberman [138] provide a useful survey of agricultural adoption in developing countries.

innovators and the contraction of laggards can be major factors in the spread of a new technology (see Metcalfe [332] for example).

Many early adoption studies used simple statistical techniques to investigate relationships.[126] For example, Gross [196] studied attitudes toward hybrid rice adoption, finding that 42% of farmers thought profitability was of primary importance, 38% thought farmer-specific factors such as experience and education to be most important, and the remaining 20% thought credit availability predominated. Hypothesis tests based on chi-square contingency tables were used extensively, for example by Wilkening [563], to investigate the effect of family decision-making processes on the adoption of hybrid corn in the United States. Although these tests establish that the relationship between the variables is (or is not) statistically significant, the relation is not quantified. Various simple correlation techniques have been applied to a wide range of problems. For example, Rogers [420] found a positive relationship between farmer contact with the agricultural extension agent and adoption, for a United States sample. Williams [566] performed a similar operation using Nigerian data. Both found contact with the extension agent to the more important than the mass media. These few examples are drawn from a sample of several hundred, more than 460 of which are listed in an early survey article by Jones [246], to which the reader is referred. Less common methods include factor analysis, used by Greene [185], who found that 6 out of 53 explanatory variables explained 52% of the variation in quantity of fertilizer used by Thai farmers; and discriminant analysis, used to classify observations in one category or another using several explanatory variables [578].

Not surprisingly, the most common approach to determining the quantitative importance of various explanatory variables has been simple regression analysis, which often attempts only to explain adoption versus non-adoption rather than the extent or intensity of use of an innovation.[127] Unfortunately, ordinary least squares estimation of equations with a dichotomous or otherwise limited

[126] These statistical methods commonly used by social scientists are described in Yeomans [579, Ch. 6].

[127] Several of the huge number of agricultural development studies of the adoption of new techniques, modern inputs, and improved seed varieties are discussed in Feder, Just, and Zilberman's survey [138].

dependent variable is not appropriate since the error structure is heteroscedastic; the parameter estimates are inefficient. Nor can classical hypothesis tests be applied since the error terms for a limited dependent variable will not be normally distributed [10, 305, 396]. Additionally, if a dichotomous variable is to be explained by the exogenous values of an attribute, it is convenient to be able to interpret the expected value of the dependent variable as the probability of adoption. For the simple linear probability model, *predicted* values of the endogenous variable may well lie outside the interval (0, 1), which violates the probabilistic interpretation [396].

The solution is to apply a transformation that will ensure that all values of the dependent variable lie in the (0, 1) interval. The obvious candidate is the cumulative probability function, which may be written,

$$P_i = F(\alpha + \beta X_i) = F(Z_i) \tag{38}$$

where F represents the cumulative function and X is stochastic. If the index Z_i is assumed to be uniformly distributed, the result is the constrained linear probability model. If Z_i is taken to be normally distributed, the result is the probit model, and if F is taken to be the cumulative logistic curve (with which this survey began), the result is the logit model.

Agricultural adoption studies abound, frequently mixing economic and "sociological" explanatory variables. Hill and Kau [230] applied the probit model to the use of corn dryers in U.S. agriculture. Feder and Slade [141] have fitted a logit model to the adoption of new techniques by Indian rice farmers. Jamison and Lau [240] used a logit model to analyze the adoption of chemical inputs by Thai farmers and found that education, age, and extension activity all had positive effects on adoption. Rahm and Huffman [406] applied a similar model to the adoption of reduced tillage by Iowa corn farmers. Gerhart [166] applied probit analysis to explain adoption rates of hybrid corn in three regions of Kenya. His use of the presence of drought-resistant crops as an indication of high risks raises a further econometric issue. Feder, Just, and Zilberman [138] point out that the planting of drought-resistant crops is itself an endogenous variable. Thus, its inclusion as an independent variable raises the issue of simultaneous equation bias, which also arises when the adoption of improved seed varieties is explained by fertilizer use, where these are really simultaneous decisions. The logit analysis of the adoption of several innovations in Philippine agriculture by Nerlove and Press [363] is a pioneering attempt at dealing with interactions of this nature.

Dichotomous choice models do not explain the *intensity of use* of the innovation, which may often be of greater importance [488]. Many studies have investigated the *intensity* of adoption by creating a dependent variable with values in the interval (0, 100), which may be interpreted as the percentage of adoption" [138, p. 285]. Specification difficulties include avoiding predictions that fall outside the interval and treatment of truncated variables, such as those that cannot take negative values. Feder, Just, and Zilberman [138] suggest a two-stage procedure for problems such as fertilizer use. A dichotomous choice model could be used to determine the probability of fertilizer use; then, given adoption, the level could be explained by a conditional model with the log of fertilizer use as the dependent variable. Alternatively, the Tobit model can estimate both the probability of adoption and the intensity of use. It has been used by Akinola [5] to explain chemical input use levels on Nigerian cocoa farms and by Shakya and Flinn [490] for fertilizer intensity in rice production in the Nepal Teroi.

Lastly, if knowledge of technologies depends on experience, the new must be more uncertain than the old, and the adoption decision will depend on the potential adopter's attitude toward risk. The literature on risk in agricultural economics spans several decades [13]. Relatively recently it has been applied to the problem of new technology, producing models that study framers' optimization problems (profit or utility maximization) as an allocation decision involving a traditional technology and an uncertain modern alternative requiring commercial inputs such as fertilizer, pesticides, and irrigation [138].

Hiebert [229] investigates the effect of imperfect information (represented by a random element in the effect of fertilizer on yields) on the adoption of a modern production technique. He finds that the risk-averse farmer uses less land and fertilizer in modern production than one who is risk-neutral. Though his model is static, Hiebert reasons that learning shifts the conditional distribution of net income from the modern technique, making adoption more likely. Feder [136] analyzes the effects of risk, risk aversion, farm size, and credit constraints on input use, output scale, and crop mix decisions, using for the modern process a stochastic production function suggested by Just and Pope [253]. With no credit constraint, the level of fertilizer per acre using modern techniques was not affected by risk aversion, uncertainty, or farm size, but greater risk resulted in a smaller allocation of land to the modern technique.

Just and Zilberman [254] extend the analysis to encompass all inputs and included the covariance of net income per hectare under modern and traditional techniques, which proves to be an important determinant of adoption intensity. Feder [137] discusses the introduction of two explicitly interrelated innovations. Feder and O'Mara [139] and Just, Zilberman, and Rausser [255] incorporate fixed transaction costs and information acquisition costs for the new technology, leading to the crucial result that farms below a critical level of size will not adopt the new technology. Returns to scale may prevail in adoption even if the modern technology itself is scale-neutral.

The predictions of these models are dependent to some extent on their initial assumptions, which include concave and well-behaved utility functions. Feder, Just, and Zilberman [138] report that rather different results are obtained from "safety first" models, in which the utility of income is assumed to be zero below a "disaster level" and unity above it [33, 403, 440].

3.5. Theoretical developments

Although the disequilibrium "epidemic" model has frequently produced good empirical results and may adequately describe the spread of diseases, fashions, and gossip [105, p. 10], critics such as Stoneman [522] stress its lack of economic content and doubt its general relevance.

3.5.1. Threshold of probit models. The static probit and logit models, introduced above in a statistical context, make adoption a function of the characteristics of adopters at one point in time. If the exogenous "stimulus" variables that explain adoption *change over time,* then an increasing proportion of the population will cross the "threshold" and adopt. The result of these dynamic "probit" models is a diffusion curve that is a function of the actual explanatory variables, rather than a function of some proxy variables such as time. These are *equilibrium* models, since the system has adjusted to the particular values of the variables at each point in time, rather than being out of equilibrium and approaching a distant final ceiling level of diffusion.

Models of this type have only recently been applied to process innovations but have a long history in the study of the diffusion of consumer durables. Pyatt's [402] impressive contribution to this area includes a brief survey of earlier work. Bain's [23] study of television ownership, which begins with a useful survey, argues that the actual growth curve will be an envelope of short-run lognormal diffusion curves (see the discussion under "Dynamic Models," Section 3.33). Cramer [95], himself a pioneer in this area, provides an introduction to the theory and estimation of lognormal Engel curves. The basic approach can be illustrated by reference to Bonus [66]. Suppose that the income of household i at time t is lognormally distributed

$$y_{it} \sim \Lambda(M_t, \sigma_t^2) \tag{39}$$

and so is the "critical level" of income ($\bar{y}_{it}$) required for adoption to occur:

$$\bar{y}_{it} \sim \Lambda(\bar{M}_t, \bar{\sigma}_t^2). \tag{40}$$

It follows that the probability that the household will own the product (P_{it}) is given by $P_r(\bar{y}_{it} \leq y_{it})$. Thus, the probability of ownership will be related to income by a "quasi-Engel curve" that will be cumulative lognormal, assigning "at a given point of time and to each income level, the corresponding fraction of actual owners" [66, p. 657]. Aggregating the quasi-Engel curve over the income distribution for each time period generates the diffusion curve that results as the distribution of income levels and adoption thresholds both change over time. Recent examples of applications of this "probit" model include Dagenais [97] (automobiles) and Wilton and Pessemier [570] (electric vehicles).

The rationale of probit (or logit) models as a description of the diffusion process has been stated by David [99]. "Whenever or wherever some stimulus variate takes on a value exceeding a critical level, the subject of the stimulation responds by instantly determining to adopt the innovation in question. The reason such decisions are not arrived at simultaneously by the entire population of potential adopters lies in the fact that at any given point of time either the 'stimulus variate' or the 'critical level' required to elicit an adoption is described by a distribution of values, and not a unique value appropriate to all members of the population. Hence, at any point in time following the advent of an innovation, the critical response level has been surpassed only in the cases of some among the whole population of potential adopters. Through some exogenous or endogenous process, however, the relative position of stimulus variate and critical response level are altered as time passes, bringing a growing proportion of the population across the 'threshold' into the group of actual users of the innovation."

This approach allowed David [101] to offer an appealing explanation of the 20-year lag between Obed Hussey's first sale of his mechanical reaper and the first wave of popular acceptance in the mid-1850s. The "stimulus variate" is farm size (S), since adoption will take place only if the saving in wages due to the reduction in labor use exceeds the cost of the reaper. Thus adoption will be profitable for farm i at time t if,

$$w_t(L_{it}^{\mathrm{O}} - L_{it}^{\mathrm{N}}) > P_{it}^{\mathrm{N}} \tag{41}$$

where w_t is the wage rate, P_{it}^{N} is the annual cost of the mechanical reaper, and L_{it}^{O} and L_{it}^{N} are the annual labor requirements for the old and new techniques respectively.[128] Let

$$L_{it}^{\mathrm{O}} = a_1 S_{it} \tag{42}$$

and

$$L_{it}^{\mathrm{N}} = a_2 S_{it}, \tag{43}$$

where the coefficients a_1 and a_2 are determined by the technology and S_{it} is farm size. Substitution gives the result

$$S_{it} > \frac{P_{it}^{\mathrm{N}}}{w_t}\left(\frac{1}{a_1 - a_2}\right), \tag{44}$$

which suggests that diffusion will occur if *either* S_{it} increases, or if the wage rate rises relative to the price of the reaper (both old and new technologies remaining unchanged). Figure 3.2 shows the relative frequency distribution of farm size. If S_{it}^* is the "critical

[128] This formulation is from Davies [105, pp. 30–31].

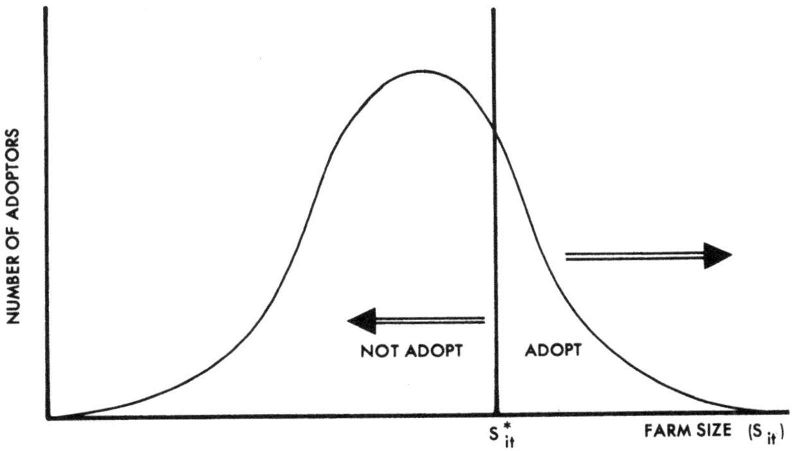

FIGURE 3.2 Relative frequency distribution of farm size.

level" above which the reaper is adopted at time t, an increase in wages relative to the cost of the reaper will shift S_{it}^* to the left[129] as time passes. Simultaneously, the distribution of farm size moves to the right. Thus a sigmoid diffusion curve is generated.

David's approach has been criticized by Olmstead [380, p. 328], who argues that "if farmers could share or rent reapers and mowers, then the threshold argument, as presently constituted, is rendered inoperative." Olmstead also stresses that there were numerous small improvements to the reaper that gradually raised its productivity (i.e. a_2 in Eq. (44) was not constant).

The model developed by Davies [105] is similar, but has the advantage of taking post-innovation technical improvements into account. He suggests that firm i will adopt by time t if

$$E(R_{it}) < R_{it}^* \tag{45}$$

where $E(R_{it})$ is the expected payoff period and R_{it}^* is the "critical level"—the maximum payoff period that firm i finds acceptable. Again, firm size is emphasized, with both the expected payoff and the critical level being defined as multiplicative functions of firm

[129] Note that this is consistent with the notion of induced mechanical innovation in U.S. agriculture studied in Part 2 of this survey.

size, other factors representing the *technical attributes of the firm*, and an error term. Both the error structure and firm size are assumed to be lognormally distributed.

The *characteristics of the innovation* are modeled by considering two types of new technology. Group A innovations are reasonably cheap and simple, while those in group B are expensive and complex. Thus post-innovation improvements decline for group A during the diffusion period as do the returns from information search, whereas for group B these factors remain constant. This assumption results from the difference between the learning possibilities for simple and complex innovations [106, p. 158]. This difference is modeled by defining the diffusion curve[130] for group B innovations to be the symmetric cumulative normal, while that for group A is the positively skewed cumulative lognormal. The slope of the curves, or speeds of diffusion, will depend on the variances of the distributions. The parameters representing the speeds of diffusion are b_1 and b_2 in the estimating equations,

$$\frac{n_t}{N} = a_1 + b_1 \log t \tag{46}$$

$$\frac{n_t}{N} = a_2 + b_2 t, \tag{47}$$

where n_t/N is the proportion of the population that has adopted. Applied by Davies to 22 innovations (8 classified as group A, 7 in group B, and 7 "unclassified"), the model appeared to be consistent with the data, tending to perform better than the logistic curve alone and giving poorer results when the "wrong" curve was fitted for the classified innovations. The sign of b_i was positive and this parameter was significant in all cases, meaning that larger firms do adopt more rapidly. This is consistent with Davies' reasoning, which stressed returns to scale.

A second-stage analysis then attempts to explain differences in

---

[130] Davies [105, pp. 75–80] fully explains the relationship between the probability of *adoption* (P_i in Eq. (38)) and the *diffusion* curve. "Conceptually the link between P_{it} and n_t/N is straightforward. n_t/N is simply the weighted sum of all possible values of P_{it}, where the weights are the probabilities that each firm size will actually occur." In other words, to derive n_t/N, the P_{it} must be integrated over the firm size distribution.

the speeds of diffusion, b_1 and b_2. The parameters are interpreted to suggest that diffusion will be faster:

(i) the greater the growth rate of the industry;
(ii) the greater the profitability of the innovation;
(iii) the greater the labor intensity of the industry;
(iv) the more important are post-invention improvements;
(v) the more effective is information search;
(vi) the smaller are inequalities in firm size; and
(vii) the smaller are inter-firm differences in expected profitability.

Conceptually, the diffusion mechanisms may be viewed as analogous to the David model schematized in Figure 3.2. Thus, as firm size may be expected to increase growth, (i) will shift the frequency distribution of *firm sizes to the right.* Factors such as post-invention improvements (iv) and information search (v) will shift the "critical value" R^*_{it} to the left. However, R^*_{it} itself has a frequency distribution, rather than being a single value, so that the diffusion process rests on the interaction of these two frequency distributions, moving in opposite directions.

Variations on the David/Davies approach can be found in Von Tunzleman's [554] analysis of the diffusion of steam power (reported quite thoroughly by Stoneman [522, Ch. 10]), and in Gutkind and Zilberman [200], whose model has a fixed frequency distribution for firm size. Their sigmoid diffusion curve results from the falling capital cost of the innovation and from increased profitability, due to learning by doing, both of which lower the adoption threshold.

Davies' model has been described in some detail, since it does represent a major advance, but several criticisms remain. First, though post-innovation improvements are attributed to learning by doing and information search is frequently mentioned, the learning process is not explicitly modeled. Second, the model concentrates on demand-side phenomena, with the diffusion process in part being driven by exogenous changes in factor prices. These are endogenized in Section 3.6, where the supply of the innovation is discussed.

3.5.2. Learning models. Stoneman [522, p. 76] considers the "driving force" of Mansfield's [314] approach to intra-firm diffusion to be the reduction in perceived risk resulting from usage. Learning by doing, or more accurately, learning by using (Rosenberg [437, Ch. 6], plays an uncertainty-reducing role in many diffusion models.

An early attempt at explicitly modeling the learning process is the study by Kislev and Shchori–Bachrach [277] of agricultural innovations. A knowledge function in the production relationship depends on both initial levels of skill and "Arrow-type learning by doing." The more highly skilled producers are more efficient in acquiring knowledge and are early adopters, but because learning by doing is communal (dependent on the industry's cumulative aggregate output), the less skilled can over time acquire sufficient knowledge to adopt.[131] As adoption spreads, the increase in supply depresses the price of the product and the more skilled, whose labour has a higher opportunity cost, are driven out. They move on, adopting the next new product or new technique, thereby generating an "innovation cycle" that owes something to both Cochrane's [91, Ch. 19] "technology treadmill" and Vernon's [551] "product cycle" in international trade. A similar "learning by doing" process relaxes the farmer's subjective constraints that limit adoption of new techniques in the programming models of Day and Singh [108] and in Feder and O'Mara's [140] simulations that lead to a familiar S-shaped diffusion curve. Feder and Slade [141] have further distinguished between the passive acquisition of information and active accumulation, which entails costs.

Though these models produce sensible results that appear to be empirically supported, the relationship between learning, decision-makers' uncertainty regarding production parameters, and the actual adoption decision is not made explicit. Stoneman and Ochoro [526] apply a mean-variance approach to intra-firm diffusion, showing that different learning processes produce different diffusion paths, but again without justifying any of the learning mechanisms. Several promising models allow experiences with the new technique to yield sample information that decision-makers use to adjust their subjective probabilities, consistent with the spirit of Bayes' theorem. Thus, in adoption studies such as O'Mara [382], prior beliefs are modified on the basis of observed performance, gener-

[131] That there are externalities among adopters in models of this type has attracted little interest, except for Allen's [8] application of Markov random field models.

ating a Bayesian posterior distribution. Feder and O'Mara [140] use Bayesian learning to justify the inclusion of the cumulative use of the innovation in the adoption function. This relationship was included by assumption in earlier studies such as Kislev and Shchori–Bachrach [277].

Pursuing the Bayesian approach, Lindner, Fischer, and Pardey [297] have studied the "innovation assessment lag," which is defined to be the passage of time between initial awareness of the innovation and actual use. In year zero, the risk-neutral farmer has a negative expectation of the normally distributed mean profit from using the innovation. A limited amount of information is collected in each period and the adoption decision is based on accumulated knowledge. Under this formulation, the innovation lag will be lengthened by greater variance of actual profit. Fischer and Lindner [158] have extended the model to allow for differences between farms, while Lindner [296] has shown that even if the innovation is scale-neutral, larger farms will adopt more quickly. This result has far-reaching implications and may be viewed as an extension of Nordhaus' [366] observation that perfect competition is incompatible with a system in which firms undertake their own R & D. Even if innovations are supplied free of charge by the public sector, the ability of larger firms to spread learning costs over a greater output will have the same consequence. The effect of *farm size* on the adoption of modern high-yielding seed varieties is crucial to appraising the effects of the green revolution. The considerable empirical literature, surveyed in some detail by Feder, Just, and Zilberman [138], suggests that larger farms do adopt more quickly. This result appears in one of the seven "generalizations" in Ruttan's [448] summary of the green revolution experience.

Bayesian learning processes have been applied to the diffusion of non-agricultural techniques by Stoneman [521] and Jensen [243] for the intra-firm case, and by the same authors (Stoneman [520], Jensen [242]) for the inter-firm situation.[132] Stoneman [521] develops the mean variance approach of Stoneman and Ochoro [526] by adding a Bayesian theory of learning and adjustment costs. The

[132] Stoneman's work is emphasised here partly because his textbook account [522] is accessible to a wider readership than Jensen's elegant papers and partly because his approach provides insights on the earlier literature.

interaction of learning and the procedure for choice of technique, plus adjustment costs, generates a diffusion curve. Let α_t be the proportion of the firm's output produced by the new technique and α_t^* be the desired level, with returns to both old and new technologies perceived to be normally distributed.

$$\text{New:} \quad N(\mu_{nt}, \sigma_{nt}^2) \tag{48}$$

$$\text{Old:} \quad N(\mu_{ot}, \sigma_{ot}^2). \tag{49}$$

Returns are additive, so at time t the firm will have a mean and variance of actual returns (following Stoneman's notation),

$$\mu_t = \alpha_t \mu_{nt} + (1 - \alpha_t)\mu_{ot} \tag{50}$$

and

$$\sigma_t^2 = \alpha_t^2 \sigma_{nt}^2 + (1 - \alpha_t)^2 \sigma_{ot}^2 + 2\alpha_t(1 - \alpha_t)\sigma_{not}, \tag{51}$$

where σ_{not} is the covariance term. α_t is chosen by maximizing a utility function subject to an adjustment cost constraint. With no adjustment costs, the solution for the desired value α_t^* is a function of the mean and variance terms in Eqs. (50) and (51) and a parameter representing the firm's attitude toward risk. Then, adjustment costs are defined to be an increasing function of the rate of change of α_t and a decreasing function of the starting level α_{t-1}. Maximization of utility subject to this cost function leads to the equation,

$$\frac{d\alpha_t}{d_t}\frac{1}{\alpha_t} = \frac{1}{\theta}\left\{\mu_{nt} - \mu_{ot} + b(\sigma_{ot}^2 - \sigma_{not})\left(\frac{\alpha_t^* - \alpha_t}{\alpha_t^*}\right)\right\}. \tag{52}$$

The parameter θ represents the adjustment cost, b is the firm's "risk coefficient," and $\sigma_{not} \equiv \rho\sigma_{ot}\sigma_{nt}$ is the covariance, which may also be expressed in terms of the correlation coefficient ρ.[133]

Though the parameters of the old technology are known with certainty, Bayesian learning is assumed to change the anticipated values of α_t and α_t^*, where the time path of α_t is the diffusion curve and α_t^* is the final level.

---

[133] If $b > 0$, then the firm is risk averse. The implication of the term containing b is that if the variance of the new technique is less than that of the old and/or if the old and new techniques are less than perfectly correlated, then diffusion of the new technique will reduce risk and will be faster for a risk averse firm than for one that is risk neutral.

The approach of α_t to the ceiling value follows the logistic curve in the special case where there is no learning and diffusion is governed by adjustment costs. This result goes some way toward integrating investment theory and the diffusion literature. When learning is included, the model can generate a sigmoid diffusion curve and can explain the failure of some innovations, since although the estimate of the variance of the new technique falls with learning, the estimated mean profitability of the new technique may either rise or fall as knowledge improves. The rate of diffusion will be greater, the higher is the true profitability of the innovation, but will also be influenced by attitudes to risk (b), the initial uncertainty (σ^2_{nt}), adjustment costs (θ), and the covariance between returns to the old and new techniques (ρ).

By way of contrast with this broad approach, Jensen's [243] theoretical paper proves that differences in prior beliefs among firms are sufficient to generate a sigmoid diffusion curve even in the absence of external information derived from the activities of other firms. Jensen's [242] decision theoretic model of adoption and inter-firm diffusion reaches the same conclusion, that "firms will adopt at different dates if and only if their original beliefs differ." Reinganum [415] uses a similar model. Stoneman's [520] inter-firm model follows a methodology similar to Stoneman [521], leading to a threshold model of the adoption decision. The assumption that the mean and variance of returns to the innovation are lognormally distributed gives a cumulative log-normal diffusion curve.

3.5.3. The game theoretic approach.[134] Jensen's result, that differences in prior beliefs are sufficient to generate a diffusion curve, contrasts with the classical diffusion model in which the process results from asymmetric information and with threshold models that rely on physical differences between firms. Recently, the problem of the timing of adoption in duopoly models, addressed earlier by Scherer [476] and Rao and Rutenberg [408], has attracted renewed attention.

Two contributions by Reinganum [413, 414] demonstrate that even if firms are identical and information on a capital-embodied

[134] The meaning of the terms *decision theoretic* and *game theoretic* and the way in which these models have been applied in the study of innovation are covered by Kamien and Schwartz [263, pp. 105–108].

innovation is perfect, strategic behavior alone can lead to a Nash equilibrium of different adoption dates, and hence a diffusion curve. This is demonstrated in a duopoly game [413], or an oligopoly game [414] that also shows that an increase in the number of firms can delay adoption. Fundenberg and Tirole [164] argue that Reinganum's model rests on the assumption that firms must precommit themselves to adoption dates. This rules out preemptive behavior, which is studied in their own model.[135]

3.6. The supply of new products

Since economists tend to rely on the interaction of supply and demand in the solution of most basic problems, it is odd that the diffusion theories discussed above pay so little explicit attention to the supply of new products in which innovations are embodied. The models considered thus far concentrate on the profitability of the innovation to the user and pay little attention to the supply side. Thus, the supply of innovations appears in Griliches' [190] early study as the "date of origin" of the diffusion process. This was taken to depend on availability of seeds, which was in turn explained by profitability to seed producers. In other studies, such as David [101], the supply side appears in exogenous price and quality changes that affect the diffusion process but are not explicitly modeled.

3.6.1. Product innovation. As Blaug [61] pointed out, the product innovations of the machine-goods industries are the process innovations of the consumer-goods industries. Thus the diffusion problem cannot be described effectively without attention to the behavior of the suppliers of the innovation as well as the users. An early move in this direction of including the behavior of suppliers is found in Glaister's [172] study of optimal advertizing policy for new consumer goods. Glaister follows epidemic model arguments to justify the derivation of a "logistic" equation that differs in one crucial point: the speed of diffusion coefficient β is a function of the

[135] An unpublished survey by Stoneman [523] offers further interpretation of the models considered in Section 3.5, which are updated to include several developments still at the discussion paper stage.

product price. If β depends upon price and the price is chosen by a monopoly supplier (with constant unit costs and a constant price elasticity) in order to maximize future receipts, then at any instant in time the firm chooses from a whole family of logistic curves. The resulting diffusion path is not logistic, but positively skewed. This is the same outcome as the "generalized static model" that also arose from a combination of diffusion by word of mouth and from a constant source (i.e., advertizing), but the route by which Glaister reaches the result is different.[136]

Bass [29] provides a more fully developed version of the generalized static demand-side model [28], which incorporates an industry supply curve that falls over time due to learning by doing.[137] Thus, the suppliers' marginal cost for the qth unit at time t is,

$$\frac{\partial C_t}{\partial q_t} = kE_t^{-\lambda}, \quad \lambda > 0 \tag{53}$$

where E_t represents the cumulative output. The demand relationship is

$$q_t = f(t)\gamma p_t^{-n} \tag{54}$$

where q_t is output, n is a constant elasticity, and $f(t)$ shifts the demand curve over time. Choosing price so as to maximize profit subject to the constant demand elasticity gives

$$p_t = \left(\frac{n}{n-1}\right)kE_t^{-\lambda}. \tag{55}$$

Substitution of (55) into (54) and manipulation lead to the expression

$$q_t = \left(\frac{m}{1-\lambda n}\right)f(t)F(t)^{\lambda n/(1-\lambda n)}, \tag{56}$$

where m is the final level of sales and $F(t)$ is the integral of $f(t)$.

---

[136] There is also common ground with Bain [23], whose skewed curve is the envelope of short-run lognormal curves and the other envelope approaches discussed in section 3.3.3.

[137] Comments on the Bass model and a summary are provided by Horsky [232], while Russell [443] provides a diagrammatic derivation of the diffusion curve.

Then $f(t)$ is defined to be the differential equation of the generalized linear model,[138]

$$f(t) = B_0 + (B_1 - B_0)F(t) - B_1(F(t))^2. \qquad (57)$$

Substitution of expressions for $f(t)$ and the solution for $F(t)$ into (56) lead to a diffusion model in which learning by doing lowers costs and hence price, while the demand side diffusion process simultaneously shifts the demand curve. The overall diffusion curve (the path of q_t) thus depends on both the shifting of demand and the effect of the falling supply price.

Several other studies have emphasized the behavior of the industry supplying a product innovation, paying little attention to the demand side or the diffusion curve. Spence [509] considers the strategic interaction among firms during the growth phase of a new industry. In a model that does not take account of the effects of learning by doing or of uncertainty, he finds that firms invest in capital equipment as rapidly as possible up to some target level and then stop. This result occurs quite generally because a firm that gets ahead can preempt the market, increasing its share and deterring entrants. A learning curve is added in Spence [510], with unit costs depending on accumulated output. This gives an advantage to early entrants and creates barriers to entry. More entry occurs if the learning effects spill over, rather than being firm-specific, and the industry growth rate is increased when demand-side learning is added to the model. The interesting point is that diffusion is not the object of attention. The aim, following in the footsteps in Schumpeter [485], is to develop "a model of competitive interaction and industry evolution." Neo-Schumpeterian models of this type, for oligopolistic industries, now constitute a literature that is quite distinct from the study of diffusion in competitive industries such as agriculture.

In a similar vein, Gort and Konakayama [182] examine "diffusion in the production of an innovation." This is defined as the increase in the number of producers of a new product (i.e. cumulative entry, less exit). In a study of seven innovations, entry was found to depend on the demonstration effect, technical change, dynamic adjustment costs, and the growth of transferable experience (i.e., as

[138] This is Eq. (33), with $n_t = F(t)$, $N = 1$, and $a_t = f(t)$.

old-firm personnel move to new entrants). Exiting firms apparently had poor endowments of intangible capital (measured by patent rate and the accumulated stock of experience of producers). Using the same definition of diffusion, Gort and Klepper [181] investigated the evolution of forty-six new product markets. They found a positive correlation between entry and technical innovation. In a manner reminiscent of Kuznets [281], they identified five stages in the evolution of the industries. These are

(i) first commercial use, up to,
(ii) period of sharp increase in the number of producers
(iii) period with a net entry of approximately zero
(iv) period of negative net entry
(v) a second period of net entry of approximately zero.

Their evidence suggests that this "product cycle" is clearly related to product innovation, with the growth phase (ii) corresponding to rapid rates of innovation emanating from firms *outside* the industry.

3.6.2. Process innovation. The contribution of Metcalfe is decidedly less neoclassical than the works discussed above. Metcalfe [332, 333, 335] develops a neo-Schumpeterian model that is explicitly intended to broaden diffusion analysis to encompass theories of industrial growth and structural change.[139] With this aim, Metcalfe surveys the contributions of Schumpeter [487], Burns [76], and Kuznets [281, 282] on industrial growth and retardation, Hicks [228, Ch. 2] on innovation-induced impulses to economic growth, and Pasinetti [390] on structural change. The standard diffusion model does not attempt to incorporate economic contributions of this type.

To close the gap, Metcalfe specifies a logistic relationship for the proportional growth of demand,

$$\frac{dq}{dt}\frac{1}{q} = B(m(p) - q_t). \tag{58}$$

[139] In evolutionary or neo-Schumpeterian models, technological progress depends on both innovation and imitation. Iwai [238] does not model the supply of innovations but does use an array of logistic curves to describe the imitation process and evolution of the industry with co-existing production methods. A second paper [239] extends the analysis to consider the evolution of technology under the combined pressures of imitation, innovation, and economic selection.

The model is dynamic in that the equilibrium demand, $m(p)$, is a function of price and may also be affected by post-innovation improvements. For simplicity,

$$m(p) = c - ap_t. \tag{59}$$

On the supply side, the rate of growth of production capacity is taken to depend on the rate of profit (as in Cambridge growth models). Hence,

$$r_t = \frac{p_t - w_t l - \gamma v}{v}, \tag{60}$$

where r is the ratio of profits to capital, v is the capital-output ratio, l is a unit input requirements coefficient, w_t is the price of a composite input, and γv is depreciation. The price of the composite input increases with output (x_t),[140] so

$$w_t = w_0 + w_1 x_t \tag{61}$$

and capacity growth is a function of the rate of profit,

$$\frac{dx_t}{dt}\frac{1}{x_t} = \theta(1 + \mu)r_t, \tag{62}$$

where θ is the fraction of profits reinvested and μ is the fixed ratio of external to internal funds invested.

Combining the last three equations and simplifying the notation gives

$$\frac{dx_t}{dt}\frac{1}{x} = \frac{p_t - h_0 - h_1 x_t}{k}. \tag{63}$$

If there are no post-innovation improvements, the growth of supply will also be logistic, with the saturation level being reached when the prime cost terms in h are equal to the output price p_t.

The role of price in equilibrating supply and demand is apparent, since the saturation levels of both equilibrium demand $(m(p))$ and the supply of productive capacity depend on the innovation's price.

[140] Stoneman [522, pp. 126–127] suggests that this proposition is not supported by the empirical evidence. For major innovations, such as the railways, Hicks [228, Ch. 2] does argue that the rate at which building occurs will be limited by scarcity of factors such as skilled labour and capital.

If p_t were too low, the growth of demand would exceed the growth of supply which would limit the rate of diffusion in a closed economy, unless p_t rises to correct the imbalance.[141]

Metcalfe combines the demand and supply side equations and solves for the balanced diffusion path of output, which is found to be logistic when there are no post-innovation improvements. When these are included, the saturation level will shift upwards over time yielding a positively skewed curve. The industry's *rate* of growth drops asymtotically toward zero (the retardation result) as profits fall because of rising costs and a falling innovation price. In keeping with Schumpeter's views, innovators make only a temporary profit. Profit is, "at the same time the child and the victim of development" [485, p. 154].

Though Metcalfe succeeds in generating several results in the Schumpeterian spirit, his model is deficient on the supply side, having only a representative firm and only one innovation. A model of Schumpetrian creative destruction and industrial evolution must allow for differences and competition between both firms and industries. A step in this direction has been taken by Metcalfe and Gibbons [336], who modified the earlier model to let two new industries appear as a result of innovations. The industries compete for resources and markets, giving rise to possible trajectories in which the growth of one industry occurs at the cost of decline in the other. Thus, potential patterns of structural change and industrial evolution are examined.

The broad scope of Metcalfe and Gibbons' approach prevents them from paying detailed attention to the modeling of the Schumpeterian firm's decision rules. Stoneman and Ireland [525] proceed in this direction with a model that adds a monopoly supplier subject to learning economies to a threshold model of the David [99]/Davies [105] type. Rather than being driven by exogenous changes in prices and firm size, the model makes the innovation's price ((P^N_{it}) in equation (44)) endogenous. Since the supplier maximizes discounted returns, early profits are attractive,

[141] In an open economy, differences between supply and demand represent imports or exports, allowing separate diffusion curves for innovation demand and the supply of productive capacity. Thus, in a model developed by Metcalfe and Soete [337], diffusion forms an explanation for international trade that is logically independent of technology gap models of the Posner [399] type.

but this tendency is balanced by rising marginal costs at any moment in time and falling costs over time due to learning. The solution to this problem determines price, which declines over time. Stoneman and Ireland show that learning economies are necessary to produce a sigmoid curve. However, no exogenous forces are required to generate the diffusion process, whereas in the David and Davies models wages were rising and technology changing.

When the monopoly supplier of the capital good is replaced by oligopoly, the stock of new machines rises and the capital goods price falls. The price will be lower, the greater the number of suppliers, but the number of suppliers does not affect the speed of diffusion.

Stoneman and David [524] consider the effects of information provision and adoption subsidies in a model that includes expectations and supply considerations and integrates the epidemic and threshold approaches to diffusion. They find that information policies are effective under perfect competition but that a monopoly supplier's reaction may negate this outcome. Subsidies to adoption will increase usage in both cases.

3.7. Aspects of the international diffusion of technology

This survey has concentrated on the development of analytical models of the diffusion process. It draws on disciplines outside economics only to the extent that they contributed to the evolution of these models. However, many of the factors that have now been incorporated in formal models were originally investigated by economic historians and other scholars such as Rosenberg [442, p. 1], who have shown "a willingness to step outside of the limited intellectual boundaries of this (economic) mode of reasoning."

3.7.1. In ecomonic history. The majority of Rosenberg's contributions are available in two volumes of collected papers [432, 434] that discuss, from a historical perspective, issues such as post-innovation improvements, complementarity between innovations, the importance of (and difference between) learning by doing and learning by using, and the improvement of old technologies that the stimulation of competition from new techniques can cause. Rosenberg's work, that of historians such as Kenwood and Lougheed [273], and of critics like Rosegger [428] show that many important elements in the extremely diverse process of diffusion have not yet been included in formal models.

In an immature literature such as that on the international diffusion of technology where formal modeling is as yet limited,[142] historical studies account for a significant proportion of current knowledge. A minority of the issues raised by the historians are considered here. For instance, David [102] extends the threshold model of the adoption of agricultural machinery to the transfer of agricultural technology from America to Britain. The slow rate of diffusion is largely explained by factors not normally included in diffusion models, such as the inappropriate topology of the English landscape, the legal and institutional arrangements, and possibly the penalties of England's early start in the accumulation of agricultural machinery. Such considerations remain relevant in the international transfer of agricultural technology at the present time.

Temin [535, 536] accounts for the slow diffusion of coke smelting from British to American iron producers largely be reference to the difference in resource endowments (wood was plentiful in America) and the unsuitability of American coal. This again calls to mind the location-specific nature of innovations and the influence of factor endowments as an *inducement mechanism.* These same influences appear in the relatively rapid change from woooden to iron ships in Britain as compared with America (studied by Harley [209]) and the mechanization of gun-making (Ames and Rosenberg, [11] and Blackmore [59], both in Saul [474]). Blackmore's study of Colt's London armory also stresses that the international diffusion of techniques required the international movement of skilled labor (see footnote 112, on migration diffusion).

In the extensive literature on the entrepreneurial failure of Victorian Britain, Sandberg [465] absolved the Lancashire cotton industry of blame for the slow diffusion of ring spinning because it had less need to economize on labor. However, Lazonick [289] attributed Lancashire's decline to a failure to make the transition from competitive to corporate capitalism. The complexity of factors affecting diffusion, often missing from mathematical models, is clear in these works and is reinforced by Allen's [9] study of iron and

[142] Formal modeling of technology in trade theory and empirical tests of such relationships have been reviewed by Pugel [401] and Cheng [86].

steel. A survey of "entrepreneurial failure" in several industries is provided by Sandberg [466] and is a main topic of McCloskey's [328] conference volume.

From this brief excursion into economic history, it is difficult to avoid the conclusion that aspects of technical change such as induced innovation, the effect of market structure, appropriate technology, diffusion, and technology transfer interact in a complex manner, especially in the context of international diffusion.[143]

3.7.2. In agricultural development. The international diffusion of agricultural technology is not a recent phenomenon. Ruttan and Hayami [456] cite the classical studies of Sauer [472] and Vavilov (in Chester [87]), which document the international diffusion of plants, animals, tools, and husbandry practices. But Evenson [129] warns that the transfer of agricultural technology is far more complex than diffusion models in economics and sociology imply. The fundamental difficulty in agricultural diffusion is locational specificity, which is defined to mean that the value of a technique depends on soil, climate, and economic conditions. This limits returns to scale in agricultural research and raises adaptive research to the status of a prerequisite for diffusion.

These difficulties are illustrated by Evenson, Houck, and Ruttan's [131] prototype study of sugar cane varieties[144] and its extension by

[143] A major gap in the literature, which we have not attempted to fill, is in the implications of recent advances in our understanding of the process of technical change for international trade. In the literature on international trade, major attention has been given to the effects of productivity growth on the terms of trade, on trade patterns, and on the partitioning of the new income streams generated by productivity growth [250, pp. 73–92; 348, pp. 22–26, 46–59]. But, except for an early article by Chipman [88] and a more recent article by Hamilton and Soderstrom [207], the relationship between the theory of induced innovation and international trade theory remains almost completely unexplored. In the standard Hechscher–Ohlin–Samuelson model, differences in resource endowments are the primary determinants of trade. In the theory of induced innovation, the path of productivity growth is directed toward releasing the constraints on growth imposed by factor constraints. To the extent that technical change can release the constraints on growth resulting from inelastic factor supplies, the power of the differential factor endowment explanation for trade is weakened [250, p. 80]. And to the extend that trade can release the constraints of factor endowments on growth, the theory of induced innovation loses part of its power to explain the direction of bias in productivity growth. Yet these two bodies of literature have not yet been adequately integrated [86, p. 184].

[144] This example is a crop-biological technology. The other categories defined by Evenson are animal-biological, chemical, mechanical, and managerial.

Evenson [129]. Evenson and Binswanger [130] stressed three findings. (1) International diffusion of new varieties was related to climate and plant disease incidence with widespread diffusion occurring only after important technological advances. (2) Early attempts to develop indigenous technologies were largely unsuccessful but did lead to screening techniques that later facilitated the diffusion of varieties from Java and India. (3) After a while, indigenous research programs were successful in adapting the Javanese and Indian varieties to local conditions in a wide range of countries.

Drawing on the same earlier work, Hayami and Ruttan [222] identified three stages in the international transfer of technology. The first, called "material transfer," is the simple importation of seeds, plants, and other materials that are adapted to suit local conditions largely through trial and error by farmers. The second phase is "design transfer," characterized by the import of journals, books, and blueprints that allow the copying and domestic production, with minor modifications, of the foreign designs. The final phase is the transfer of "capacity" and scientific knowledge, entailing an indigenous research capability that can adapt foreign prototypes to local conditions and increasingly create a truly indigenous technology.

Ruttan [447] argued that *institutional* transfer and innovation are essential elements in the development of capacity, and Ruttan and Binswanger [454] discussed the relationship between induced technical change and institutional change. However, developments of such complexity cannot be expected to occur automatically or rapidly. The establishment of the international agricultural research institutes represents the single most far-reaching attempt to increase both research capacity and the rate of diffusion. The structure of the international research system, its role in adaptive research, and its relation to developed and underdeveloped country national research systems are discussed by Ruttan [452]. Biggs and Clay [47] suggest that the international system has "filled the gap in the neglected subject of foodcrops."

Hayami [214] provides the historical background with an account of the development and transfer of new rice varieties in Asia. His account is extended to include the green revolution in Ruttan and Hayami [456]. The paper by Evenson and Binswanger [130] represents a move in the direction of formal modeling, reporting the results of the Evenson and Kislev [132] empirical investigation of technical change and diffusion in cereal grains. Many studies on national and international agricultural research and productivity are drawn together in Arndt, Dalrymple, and Ruttan [14].

Though the literature has concentrated on basic technology transfer issues, a minority of authors have investigated the transfer and development of capacity in the production of particular inputs necessary to the success of the green revolution. Thus Ghatak [167] studied the transfer of fertilizer production technology and Morehouse (in Stewart and James [515]) considered the Indian tractor industry. The seed industry has been investigated by Godden [174, 175], who concentrated on the concept of plant breeders' rights. The role of multinational companies in the international supply of modern seed varieties has been raised by Mooney [343].

The international diffusion of agricultural technology raises most clearly the question of how technology transfer and the development of indigenous capacity interact. In a largely critical survey, Biggs and Caly [47] argue that the conventional wisdom on agricultural technology diffusion amounts to a center-periphery model, entailing notions of dependency. In this they follow Rogers [423], who maintains that in the classical model, the innovation originates from an expert source, such as an R & D organization, and is diffused as a uniform package to passive potential adopters. This centralized, vertical model is contrasted by Rogers [422] to a decentralized model in which innovations evolve as they diffuse. Drawing on the work of Schon [482], Rogers suggests that a decentralized model is more appropriate to situations in which users develop the innovation. A high degree of *reinvention*[145] can lead to adopters becoming their own change-agents and active participants in the horizontal dissemination of innovations.

Rogers [423, p. 346] concludes that most diffusion networks contain elements of both centralized and decentralized systems, which can be combined to give a uniquely appropriate representation. In international agricultural development, this suggests that the analysis could incorporate on-farm research and allow for

[145] This is Rogers' [423, pp. 175–184] term, "defined as the degree to which an innovation is changed or modified by a user in the process of its adoption and implementation." The crucial point is that adopters play an important role in the process rather than being merely passive recipients of the innovation.

farmer (user) participation in developing appropriate technology (CIMMYT economics staff, in Eicher and Staatz [126]). Ruttan [452, p. 135] has also argued that some elements of farming or cropping system research are essential to provide feedback to the research institutions on the technical and environmental constraints faced by farmers. A centralized/decentralized analysis would be capable of incorporating the informal, farm-level R & D network stressed by Biggs and Clay [46] and would emphasize that research should be aimed at solving farmers' problems (Biggs, in Stewart and James [515]).

A method by which these approaches can be combined arises from Rosenberg's [437] distinction between "learning by doing" and "learning by using." "Learning by doing" advances with cumulative output of the capital goods industry supplying the innovation, leading to lower costs and product prices. "Learning by using" refers to learning by the *users* of the capital good in the consumer goods industry and can take two forms. The resultant technical change may be of a disembodied nature (better results from using an unchanged innovation), or the innovation itself may need to be changed. In the second case, "what we are describing is a feedback loop" [437, p. 123], and the improvement must be embodied in the innovation by the producer. If we read "agricultural research institution" for "capital goods industry" and "small-scale farmer" for "consumer goods industry," the feedback loop is the proportion of technical change generated by informal R & D that requires embodiment in the innovation.

4. CONCLUSION

Addressing the American Economic Association in 1966, Boulding [67] chastized the profession for neglecting technical change to the point where economists were incapable of answering many of the most important questions of the day. Boulding's concern has been echoed in several subsequent reviews [263, p. 223; 354].

This review documents major advances in our understanding of the process of technical change, but a number of inadequacies remain to be resolved. Part 1 raises the issue of our inadequate understanding of how science interacts with technology. Unicausal explanations, such as the science push or demand pull models of technical change, are clearly inadequate.

This view is confirmed in the review of the literature on induced technical change in Part 2. Categorizations of commonly used concepts like factor substitution and technical change rest on an arbitrary definition of the isoquant. While the induced innovation hypothesis has met with some success in explaining the *direction* of technical change, the relationship between the *rate* of technical change, profitability, and research and development expenditures, is far less clear, as is the relationship between rate and direction of technical change.

The discussion of diffusion in Part 3 also raises definitional problems. If the distinction between (major) innovation and post-innovation improvement or re-invention is arbitrary, so too is that between innovation and adoption [423, pp. 175–182]. This is made worse by reliance on the sequence of invention-innovation-improvement that misses the point that much of the firm's technological activity is neigher acquired from nor transferred to other firms. Nor have we in this survey been able to weave the component parts into a coherent whole. The linkages between the process of cumulative synthesis and induced innovation and between induced innovation and diffusion remain unresolved.[146]

Many important problems do not fit neatly into the confines of a single discipline or even related disciplines. This is clearly true of both our attempts to understand the sources of technical change, which lie at the interface of the social and physical sciences [127], and our attempts to understand the diffusion and impact of technical change, which require a more adequate understanding of the other sources of institutional change. It seems apparant, as Kuznets [283] has argued, that we continue to inadequately capture many of the costs associated with technical change, with the result that the benefits are often exaggerated relative to the costs.

The induced innovation section began with the presumption that the tradition of treating technical change as exogneous to the economic system made inadequate use of the power that economics can bring to bear on understanding the process of technical change.

[146] Methodological problems must be expected in integrating induced innovation, a variance theory firmly based in neoclassical economics, with diffusion theory, an evolutionary or process theory [340].

A similar argument can be made in the case of institutional change, which until recently has also been treated as exogenous. Substantial progress has been made in treating institutional change as at least partially endogenous [412]. But we share with Field [157] the view that it will not be possible to endogenize fully either technical or institutional change. Both the rate and direction of technical and institutional change will be influenced by forces that are exogenous to the economic system.

We have documented the substantial progress that has been made in the attempts by economists to understand the process of technical change. But progress has been slow. The analysis of technical change involves problems such as market failure, interdependencies, historically contingent events, and the dynamics of change, which do not fit easily into the neoclassical framework. However, we do not agree with Nelson and Winter [360, p. 205] that the use of the augmented neoclassical model has led to a dead end. The power of the analytical methods and the advances in knowledge reviewed in this paper have provided too much insight into the process of technical change to accept readily the Nelson–Winter conclusion. But this should not blind us to the limitations of the neoclassical approach as we attempt to extend our knowledge. Neoclassical analysis is a "system of thought which in its pure form happens to be fundamentally ahistorical, if not actually anti-historical" [100, p. 11]. When time is dealt with, historical reality is often sacrificed to mathematical tractiability. The failure to come to grips with historically contingent events is at odds with the reality of technical change at the micro level.[147] Firms differ in their technological characteristics, in part because they have different histories and different past experiences.

By contrast, the "neo-Schumpeterian approach is concerned above all with the process of economic change, as opposed to the analysis of equilibrium states" [163, p. 609]. An increasing number of authors have now followed Nelson and Winter's lead in either recommending or contributing to the evolutionary or neo-Schumpeterian approach to technical change [127, 267, 334]. These

[147] Macro analysis of economic growth and technical change at the sector or economy-wide level has left crucial questions unanswered. The analysis must be a search for micro-foundations (the firm is the basic organism). This, Elster [127, p. 23] suggests, is "a pervasive and omnipresent feature of science."

evolutionary studies draw their inspiration from biology rather than classical physics. Social science applications of evolutionary concepts date back to Smiles.[148] In economics, Marshall [332] recommended the biological analogy before proceeding to foresake notions of life and movement for the simpler and more tractable approach of mechanical equilibrium.[149] A limited form of the evolutionary process has been applied successfully by Alchian [6] and others, but the effective application of evolutionary models to the study of technical change awaits a more rigorous development of the methodological foundations of this branch of economics.

The advances in our understanding of the role of demand and supply side forces in influencing the direction and diffusion of technical change have important implications for economic policy. The demonstration of the powerful role of economic forces in inducing technical change places a major burden on the efficiency of both market and nonmarket resource allocation systems. The theory of induced innovation and the historical research conducted within the induced innovation perspective are consistent with the inference that when either factor–factor or factor–product price relationships have been distorted, either through market or nonmarket interventions, the innovative behavior of both public research institutions and private research and development organizations will be biased.

The impact of bias in the allocation of research and development resources is particularly serious because of the long lag between the allocation of resources to research and the impact of the new technology generated by research and development on production. If rates of return to research were low, the cost of such distortions would also be low. But because the rates of return to research have been very high [319, pp. 144–146; 452, pp. 237–261], the costs of distortion are also very high. This implies that assuring the efficiency of the institutions through which resources are allocated to research and development must be a central element in policies designed to speed the process of economic growth.

The advances in our understanding of the linkages between

---

[148] See Hirschleifer [231] for a consideration of the current relationship between biology and economics.

[149] Schumpter's [486] review of Marshall's *Principles* stresses this aspect of his contribution to economics.

research and development and diffusion processes are also adding importantly to our capacity to design effective technology transfer policies. The effective diffusion of new technology is dependent on the capacity to invent and reinvent new technology. This means that a country or region that wants to acquire access to the new income streams generated by technical change must go beyond reliance on simple technology transfer and invest in the capacity to adapt the technology for its own resource and institutional environment. And when it has acquired the capacity to effectively transfer, adapt, and diffuse technology it will also have the capacity to invent technology that is appropriate to its resource endowment and institutional environment.

References

* Priority references, considered by authors to be essential to the understanding of the field. ** Publications mentioned in text but not included in first category. Publications not starred are considered by the authors to be interesting and relevant but are not referred to in the text.

[1] Abramovitz, M., "Resource and Output Trends in the United States since 1870", *American Economic Review,* **46 (1956), 5–23.

[2] Ahmad, S., "On the Theory of Induced Innovation", *Economic Journal,* **76 (1966), 344–357.

**[3] Ahmad, S. and A. Kubursi, "Induced Adjustment and the Role of Agriculture in Economic Development: A Case Study of Egypt and Syria", Department of Economics Working Paper 77-20, McMaster University, Ontario, Canada, 1977.

[4] Akino, M. and Y. Hayami, "Efficiency and Equity in Public Research: Rice Breeding in Japan's Economic Development", *American Journal of Agricultural Economics,* **57** (1975), 1–10.

**[5] Akinola, A. A., "A Theoretical and Empirical Analysis of Innovation Adoption Processes: A Case Study of Cocoa Spraying Chemicals in Nigeria", *PhD dissertation,* University of Manchester, Manchester, England, 1984.

[6] Alchian, A. A., "Uncertainty, Evolution and Economic Theory", *Journal of Political Economy,* **58 (1950), 211–221.

[7] Alderman, H., "Attributing Technological Bias to Public Goods", *Journal of Development Economics,* **14 (1984), 375–393.

[8] Allen, B., "Some Stochastic Processes of Interdependent Demand and Technological Diffusion of an Innovation Exhibiting Externalities Among Adopters", *International Economic Review,* **23 (1982), 595–608.

[9] Allen, R. C., "International Competition in Iron and Steel, 1850–1913", *Journal of Economic History,* **39 (1979), 911–937.

[10] Amemiya, T., "Qualitative Response Models: A Survey", *Journal of Economic Literature,* **19 (1981), 1483–1536.

[11] Ames, E., and N. Rosenberg, "The Enfield Arsenal in Theory and History", *Economic Journal,* **78 (1968), 730–733.
[12] Anden-Lacsina, T., and R. Barker, "The Adoption of Modern Varieties", in *Interpretive Analysis of Selected Papers from Changes in Rice Farming in Selected Areas of Asia.* Los Banos, Philippines: International Rice Research Institute, p. 33, 1978.
[13] Antle, J. M., "Incorporating Risk in Production Analysis", *American Journal of Agricultural Economics,* **65 (1983), 1099–1106.
**[14] Arndt, T., D. Dalrymple and V. Ruttan, *Resource Allocation and Productivity in National and International Agricultural Research.* Minneapolis: University of Minnesota Press, 1977.
[15] Arrow, K., "The Economic Implications of Learning by Doing", *Review of Economic Studies,* **29 (1962), 155–173.
[16] Arrow, K., "Classificatory Notes on the Production and Transmission of Technical Knowledge", *American Economic Review,* **81** (1968), 29–35.
[17] Asher, E., "Industrial Efficiency and Biased Technical Change in American and British Manufacturing: The Case of Textiles in the Nineteenth Century", *Journal of Economic History,* **32 (1972), 431–442.
[18] Atkinson, A. B. and J. E. Stiglitz, "A New View of Technological Change", *Economic Journal,* **79 (1969), 573–578.
[19] Averch, H. and L. Johnson, "Behaviour of the Firm under Regulatory Constraint", *American Economic Review,* **52 (1962), 1153–1169.
[20] Ayres, R. U. "A Schumpeterian Model of Technological Substitution", *Technological Forecasting and Social Change,* **27 (1985), 375–383.
[21] Babcock, J. N., "Adoption of Hybrid Corn: A comment", *Rural Sociology,* **27 (1962), 332–338.
[22] Bailey, E. E. and J. D. Malone, "Resource Allocation and the Regulated Firm", *Bell Journal of Economics and Management Science,* **1 (1970), 129–142.
**[23] Bain, A., *The Growth of T.V. Ownership in the U.K. since the War.* Cambridge: Cambridge University Press, 1964.
[24] Balcer, Y. and S. A. Lippman: "Technological Expectations and Adoption of Improved Technology", *Journal of Economic Theory,* **34 (1984), 292–318.
[25] Barker, R. and R. Herdt, "Equity Implications of Technology Changes", in *Interpretive Analysis of Selected Papers from Changes in Rice Farming in Selected Areas of Asia.* Los Banos, Philippines: International Rice Research Institute, 1978, pp. 83–108.
[26] Barlow, C. and S. K. Jayasuriya: "Bias Towards the Large Farm Subsector in Agricultural Research: The Case of Malaysian Rubber", *Research and Development in Agriculture,* **1 (1984), 153–164.
**[27] Bartlett, H. R., "The Development of Industrial Research in the United States", in *Research: A National Resource,* 3 vols, ed. by U.S. National Research Council, Washington, D.C., 1941, Vol. 2, pp. 19–77.
[28] Bass, F. M., "A New Product Growth Model for Consumer Durables", *Management science,* **15 (1969), 215–227.
[29] Bass, F. M., "The Relationship between Diffusion Rates, Experience Curves and Demand Elasticities for Consumer Durable Technological Innovations", *Journal of Business,* Part II, **53 (1980), 551–567.
**[30] Battelle Research Institute, *Interaction of Science and Technology in the Innovation Process: Some Case Studies.* Columbus, Ohio: Battelle Research Institute, 1973.

[31] Beckford, G., "Strategies for Agricultural Development: Comment", *Food Research Institute Studies in Agricultural Economics, Trade and Development,* **11** (1972), 149–154.
[32] Belinfante, A., "The Identification of Technical Change in the Electricity Generating Industry", Chapter 4.3 in *Production Economics: A Dual Approach to Theory and Applications,* ed. by M. Fuss and D. McFadden. Amsterdam: North-Holland, 1978.
[33] Bell, C., "Acquisition of Agricultural Technology", *Journal of Development Studies,* **9** (1972), 123–159.
[34] Benvignati, A. M., "Interfirm Adoption of Capital Goods Innovations", *Review of Economics and Statistics,* **64** (1982), 330–335.
[35] Benvignati, A. M., "The Relationship between the Origin and Diffusion of Industrial Innovation", *Economica,* **49** (1982), 313–323.
[36] Ben-Zion, U. and V. W. Ruttan, "Aggregate Demand and the Rate of Technical Change", in *Induced Innovation: Technology, Institutions and Development,* ed. by H. P. Binswanger and V. W. Ruttan, pp. 261–275. Baltimore: Johns Hopkins University Press, 1978.
[37] Bergstrom, V. and M. Melander, "Production Functions and Factor Demand Functions in Postwar Swedish Industry", *Scandinavian Journal of Economics,* **81** (1979), 534–551.
[38] Berkson, J. "A Statistically Precise and Relatively Simple Method of Estimating the Bio Assay with Quantal Response, Based on the Logistic Function", *Journal of the American Statistical Association,* **48** (1953), 565–599.
[39] Berndt, E. R. and B. C. Field, eds, *Modelling and Measuring Natural Resource Substitution.* Cambridge, Mass.: MIT Press, 1981.
[40] Berndt, E. and M. Khaled, "Parametric Productivity Measurement and Choice among Flexible Functional Forms", *Journal of Political Economy,* **87** (61) (1979), 1120–1145.
[41] Berndt, E. R. and D. O. Wood, "Technical Change, Tax Policy and the Derived Demands for Energy", Department of Economics, University of British Columbia, 1975.
[42] Bieri, J., A. de Janvry and A. Schmitz, "Agricultural Technology and the Distribution of Welfare Gains", *American Journal of Agricultural Economics,* **54** (1972), 801–808.
[43] Biggs, S., "Agricultural Research: A Review of Social Science Analysis", Discussion Paper 140, School of Development Studies, University of East Anglia, 1981.
[44] Biggs, S., "The Implications of Bureaucratic Factors for Agricultural Policy Analysis", *Bangladesh Journal of Agricultural Economics,* **4** (1981), 35–50.
[45] Biggs, S., "Institutions and Decision Making in Agricultural Research", in *The Economics of New Technology in Developing Countries,* ed. by F. Stewart and J. James, pp. 209–224. London: Frances Pinter, 1982.
[46] Biggs, S. and E. Clay, "Sources of Innovation in Agricultural Technology", *World Development,* **9** (1981), 321–336.
[47] Biggs, S. and E. Clay, "Generation and Diffusion of Agricultural Technology: A Review of Theories and Experiences", Working Paper No. 122, World Employment Programme, Technology and Employment Programme, International Labour Organisation, Geneva, August 1983.
[48] Binswanger, H. P., "The Measurement of Biased Efficiency Gains in U.S. and Japanese Agriculture to Test the Induced Innovation Hypothesis", *PhD dissertation,* North Carolina State University, Raleigh, 1973.

[49] Binswanger, H. P., "A Cost Function Approach to the Measurement of Elasticities of Factor Demand and Elasticities of Substitution", *American Journal of Agricultural Economics,* **56 (1974), 377–386.

[50] Binswanger, H. P., "A Microeconomic Approach to Induced Innovation", *Economic Journal,* **84 (1974), 940–958.

[51] Binswanger, H. P., "The Measurement of Technical Change Biases with Many Factors of Production", *American Economic Review,* **64 (1974), 964–976.

**[52] Binswanger, H. P., "Measuring the Impact of Economic Factors on the Direction of Technical Change", Chapter 25 in *Resource Allocation and Productivity,* ed. by T. M. Arndt, D. Dalrymple, and V. Ruttan. Minneapolis: University of Minnesota Press, 1977.

**[53] Binswanger, H. P., "Induced Technical Change: Evolution of Thought", Chapter 2 in *Induced Innovation,* ed. by H. Binswanger and V. Ruttan. Baltimore: Johns Hopkins University Press, 1978.

**[54] Binswanger, H. P., "The Microeconomics of Induced Technical Change", Chapter 4 in *Induced Innovation,* ed. by H. Binswanger and V. Ruttan. Baltimore: Johns Hopkins University Press, 1978.

**[55] Binswanger, H. P., "Issues in Modeling Induced Technical Change", Chapter 5 in *Induced Innovation,* ed. by H. Binswanger and V. Ruttan. Baltimore: Johns Hopkins University Press, 1978.

**[56] Binswanger, H. P., "Measured Biases of Technical Change: The United States", Chapter 7 in *Induced Innovation,* ed. by H. Binswanger and V. Ruttan. Baltimore: Johns Hopkins University Press, 1978.

*[57] Binswanger, H. and V. Ruttan, eds, *Induced Innovation: Technology, Institutions, and Development.* Baltimore: Johns Hopkins University Press, 1978.

*[58] Blackman, A. W., "The Market Dynamics of Technological Substitution", *Technological Forecasting and Social Change,* 6 (1974), 41–63.

[59] Blackmore, H. L., "Colt's London Armory", Chapter 6, pp. 171–195, in *Technical Change: The United States and Britain in the Nineteenth Century,* ed. by S. B. Saul. London: Methuen, 1970. First published in *Gun Digest,* 1958.

[60] Blackorby, C., C. A. K. Lovell and M. Thursby, "Extended Hicks Neutral Technical Change", *Economic Journal,* **86 (1976), 845–852.

**[61] Blaug, M., "A Survey of the Theory of Process-Innovation", *Economica,* 63 (1963), 13–32. Reprinted in Chapter 4 of *The Economics of Technological Change,* ed. by N. Rosenberg. Harmondsworth, England: Penguin, 1971.

**[62] Bliss, C. J., *Capital Theory and the Distribution of Income.* Amsterdam: North-Holland, 1975.

[63] Bloom, G., "A Note on Hicks' Theory of Invention", *American Economic Review,* **36** (1946), 83–96.

[64] Bloom, G., "Union Wage Pressure and Technological Discovery", *American Economic Review,* **61** (1951), 603–617.

[65] Bogue, A. G., "Changes in Mechanical and Plant Technology: The Corn Belt, 1910–1940", *Journal of Economic History,* **43** (1983), 1–25.

[66] Bonus, H., "Quasi-Engel Curves, Diffusion, and the Ownership of Major Consumer Durables", *Journal of Political Economy,* **81 (1973), 655–677.

[67] Boulding, K., "The Economics of Knowledge and the Knowledge of Economics", *American Economic Review,* **56 (1966), 1–13.

[68] Brander, L. and M. Straus, "Congruence Versus Profitability in the Diffusion of Hybrid Sorghums", *Rural Sociology,* **24 (1959), 381–383.

[69] Brewer, A., "A Three (or More) Factor Model of Growth with Induced Innovation", *Review of Economic Studies,* **42 (1975), 285–292.
**[70] Brown, L. A., *Innovation Diffusion.* New York: Methuen, 1981.
[71] Brown, L. A. and B. Lentnek, "Innovation Diffusion in Developing Economy: A Mesoscale View", *Economic Development and Cultural Change,* **21 (1973), 274–292.
**[72] Brown, M., *On the Theory and Measurement of Technological Change.* Cambridge: Cambridge University Press, 1966.
[73] Brown, M. and J. Popkin, "A Measure of Technological Change and Returns to Scale", *Review of Economics and Statistics,* **44 (1962), 402–411.
[74] Brozen, Y., "Determinants of the Direction of Technical Change", *American Economic Review, Papers and Proceedings,* **65 (1953), 288–302.
[75] Burmeister, E. and A. Dobell, *Mathematical Theories of Econimic Growth.* New York: Macmillan Co., 1970.
**[76] Burns, A. F., *Production Trends in the United States since 1870.* New York: National Bureau of Economic Research, 1934.
**[77] Bush, Vannevar, *Science: The Endless Frontier.* Washington, D.C.: Office of Scientific Research and Development, July 1945. Reprinted, with Introduction by Alan T. Waterman, National Science Foundation, Washington, D.C., July 1960.
[78] Cain, L. and D. Paterson, "Factor Biases and Technical Change in Manufacturing: The American System, 1850–1919", *Journal of Economic History,* **41 (1981), 341–360.
**[79] Capron, W. M., *Technological Change in Regulated Industries.* Washington, D.C.: Brookings Institution, 1971.
[80] Caves, R. E., *Multinational Enterprise and Economic Analysis.* Cambridge: Cambridge University Press, 1982.
[81] Caves, D. W., L. R. Christiansen and W. E. Diewert, "Multilateral Comparisons of Output, Input and Productivity using Superlative Index Numbers", *Economic Journal,* **92 (1982), 73–86.
[82] Chambers, R. G. and H. Lee, "Constrained Output Maximisation and U.S. Agriculture", *Applied Economics,* **18 (1986), 347–357.
[83] Chang, W., "The Neoclassical Theory of Technical Progress", *American Economic Review,* **60 (1970), 912–923.
[84] Chang, W., "A Two-Sector Model of Economic Growth with Induced Technical Progress", *Canadian Journal of Economics,* **3 (1970), 199–212.
[85] Chang, W., "A Model of Economic growth with Induced Bias in Technical Progress", *Review of Economic Studies,* **39 (1972), 205–221.
[86] Cheng, Leonard, "International Trade and Technology: A Brief Survey of the Recent Literature", *Weltwirtschaftliches Archiv,* **120 (1984), 165–189.
[87] Chester, K. S., *Chronica Botanica,* translation of N. I. Varilov, *The Origin, Variation, Immunity and Breeding of Cultivated Plants,* **13 (1–6) (1949–1950).
**[88] Chipman, J. S., "Induced Technical Change and Patterns of International Trade", in *The Technology Factor in International Trade,* ed. by R. Vernon, pp. 95–127. New York: Columbia University Press for National Bureau of Economic Research, 1970.
[89] Chow, G. C., "Technological Change and the Demand for Computers", *American Economic Review,* **57 (1967), 1117–1130.
[90] Christiansen, P. P., "Land Abundance and Cheap Horsepower in the Mechanisation of the Antebellum United States Economy", *Explorations in Economic History,* **18 (1981), 309–329.

**[91] Cochrane, W. W., *The Development of American Agriculture: A Historical Analysis*. Minneapolis: University of Minnesota Press, 1979.
**[92] Coleman, K. S., *Introduction to Mathematical Sociology*. London: Collier Macmillan Ltd., 1964.
[93] Conlisk, J., "A Neoclassical Growth Model with Endogenously Positioned Technical Change Frontier", *Economic Journal,* **79 (1969), 348–362.
[94] Cramer, J. S., "A Dynamic Approach to the Theory of Consumer Demand", *Review of Economic Studies,* **24** (1957), 73.
**[95] Cramer, J. S., *Empirical Econometrics*. Amsterdam: North-Holland; and New York: American Elsevier, 1971.
[96] Craven, J., "Stability in a Two-Sector Model with Induced Bias", *Economic Journal,* **83 (1973), 858–862.
[97] Dagenais, M. G., "Application of a Threshold Regression Model to Household Purchases of Automobiles", *Review of Economics and Statistics,* **57 (1975), 275–285.
[98] Daly, M. J. and P. S. Rao, "Productivity, Scale Economies, and Technical Change in Ontario Hydro", *Southern Economic Journal,* **52 (1985), 167–180.
**[99] David, P. A., "A Contribution to the Theory of Diffusion", Memorandum No. 71, Stanford Center for Research in Economic Growth, Stanford University, 1969.
**[100] David, P. A., "Labor Scarcity and the Problem of Technological Practice and Progress in Nineteenth-Century America", Chapter 1 in *Technical Choice, Innovation and Economic Growth*. Cambridge: Cambridge University Press, 1975.
*[101] David, P. A., "The Mechanisation of Reaping in the Ante-Bellum Midwest", Chapter 4 in *Technical Choice, Innovation and Economic Growth*. Cambridge: Cambridge University Press, 1975.
**[102] David, P. A., "The Landscape and the Machine: Technical Interrelatedness, Land Tenure and the Mechanisation of the Corn Harvest in Victorian Britain", Chapter 5 in *Technical Choice, Innovation and Economic Growth*. Cambridge: Cambridge University Press, 1975.
[103] David, P. A., and T. van de Klundert, "Biased Efficiency Growth and Capital-Labor Substitution in the U.S., 1899–1960", *American Economic Review,* **55 (1965), 357–394.
[104] Davidson, W. H., "Patterns of Factor-Saving Innovation in the Industrialised World", *European Economic Review,* **8 (1976), 207–217.
*[105] Davies, Stephen, *The Diffusion of Process Innovations*. Cambridge: Cambridge University Press, 1979.
**[106] Davies, S., "Diffusion, Innovation and Market Structure", Chapter 8 in *Research Development and Technological Innovation,* ed. by D. Sahal. Lexington, Mass.: D. C. Heath, 1980.
**[107] Davis, H. J., *The Theory of Econometrics*. Bloomington, Ind.: Principia Press, 1941.
**[108] Day, R. H. and I. Singh, *Economic Development as an Adaptive Process: The Green Revolution in the Indian Punjab*. Cambridge: Cambridge University Press, 1977.
[109] de Janvry, A., "A Socioeconomic Model of Induced Innovation for Argentine Agricultural Development", *Quarterly Journal of Economics,* **87 (1973), 410–425.
**[110] de Janvry, A., "Inducement of Technological and Institutional Innovations:

An Interpretative Framework", Chapter 6 in *Resource Allocation and Productivity,* ed. by T. M. Arndt, D. Dalrymple, and V. Ruttan. Minneapolis: University of Minnesota Press, 1977.

**[111] de Janvry, A., "Social Structure and Biased Technical Change in Argentine Agriculture", Chapter 11 in *Induced Innovation: Technology, Institutions, and Development,* ed. by H. Binswanger and V. Ruttan. Baltimore: Johns Hopkins University Press, 1978.

**[112] de Janvry, A., and J. J. Dethier, "The Political Economy of Rate and Bias of Technological Innovations in Agriculture", WP341 rev., Department of Agricultural and Resource Economics, University of California, Berkeley, March 1985.

**[113] Diamond, P., D. McFadden and M. Rodriguez, "Identification of the Elasticity of Substitution and the Bias of Technical Change: An Impossibility Theorem", Working Paper No. 62, University of California Institute for Business and Economic Research, 1965.

**[114] Diamond, P., D. McFadden and M. Rodriguez, "Measurement of the Elasticity of Factor Substitution and the Bias of Technical Change", in *Production Economics: A Dual Approach to Theory and Applications,* Vol. 2, ed. by M. Fuss and D. McFadden. Amsterdam: North-Holland, 1978.

[115] Diewert, W. E., "An Application of the Shephard Duality Theorem: A Generalized Leontief Production Function", *Journal of Political Economy,* **79** (1971), 481–507.

[116] Diewert, W. E., "Functional Forms for Revenue and Factor Requirements Functions", *International Economic Review,* **15 (1974), 119–130.

[117] Diewert, W. E., "Exact and Superlative Index Numbers", *Journal of Econometrics,* **4 (1976), 115–146.

[118] Dixon, R., "Hybrid Corn Revisited", *Econometrica,* **48 (1980), 1451–1461.

[119] Dodd, S. C., "Testing Message Diffusion in Controlled Experiments: Charting the Distance and Time Factors in the Interactance Hypothesis", *American Sociological Review,* **18 (1953), 410–416.

[120] Dodson, J. A. and E. Muller, "Models of New Product Diffusion through Advertising and Word-of-Mouth", *Management Science,* **24 (1978), 1568–1578.

*[121] Dogramaci, A., "Econometric Approaches to Productivity Measurement: A Brief Overview", Chapter 1 in *Developments in Econometric Analyses of Productivity,* ed. by A. Dogramaci. Boston: Kluwer Nijhoff, 1983.

**[122] Drandakis, E. M. and E. S. Phelps, "A Model of Induced Invention, Growth and Distribution", Cowles Foundation Discussion Paper No. 186, Yale University, 1965.

[123] Drandakis, E. M. and E. S. Phelps, "A Model of Induced Invention, Growth and Distribution", *Economic Journal,* **76 (1966), 823–840.

**[124] Duncan, R. C. and H. P. Binswanger, "Factor Biases and Induced Innovation in Australian Manufacturing", mimeographed, University of Minnesota, St. Paul, 1974.

[125] Duncan, R. C. and H. P. Binswanger, "Energy Sources: Substitutability and Biases in Australia", *Australian Economic Papers,* **15 (1976), 289–301.

**[126] Eicher, C. K. and J. M. Staatz, eds, *Agricultural Development in the Third World.* Baltimore: Johns Hopkins University Press, 1984.

**[127] Elster, J., "Neoclassical Theories", Chapter 4, and "Evolutionary Theories", Chapter 6, in *Explaining Technical Change.* Cambridge: Cambridge University Press, 1983.

**[128] Enos, J., "Invention and Innovation in the Petroleum Refining Industry", in *The Rate and Direction of Inventive Activity*. Princeton: Princeton University Press for the National Bureau of Economic Research, 1962.

[129] Evenson, R., "International Diffusion of Agrarian Technology", *Journal of Economic History,* **34 (1974), 51–73.

**[130] Evenson, R. and H. Binswanger, "Technology Transfer and Research Resource Allocation", Chapter 6 in *Induced Innovation: Technology, Institutions and Development,* ed. by H. Binswanger and V. Ruttan. Baltimore: Johns Hopkins University Press, 1978.

**[131] Evenson, R. E., J. P. Houck and V. W. Ruttan, "Technical Change and Agricultural Trade: Three Examples—Sugar Cane, Bananas and Rice", in *The Technology Factor in International Trade,* ed. by R. Vernon, pp. 415–480. New York: National Bureau of Economic Research, 1970.

**[132] Evenson, R. and Y. Kislev, *Agricultural Research and Productivity*. New Haven: Yale University Press, 1975.

[133] Evenson, R. and Y. Kislev, "A Stochastic Model of Applied Research", *Journal of Political Economy,* **84 (1976), 265–281.

[134] Falcon, W., "The Green Revolution: Generations of Problems", *American Journal of Agricultural Economics,* **52** (1970), 698–710.

[135] Farrell, M. J., "The Measurement of Productive Efficiency", *Journal of the Royal Statistical Society,* **120,** Pt. 3, Series A (1957), 253–281.

[136] Feder, G., "Farm Size, Risk Aversion and the Adoption of New Technology under Uncertainty", *Oxford Economic Papers,* **32 (1980), 263–283.

[137] Feder, G., "Adoption of Interrelated Agricultural Innovations: Complementarity and the Impacts of Risk, Scale and Credit", *American Journal of Agricultural Economics,* **64 (1982), 94–101.

*[138] Feder, G., R. Just and D. Zilberman, "Adoption of Agricultural Innovations in Developing Countries: A Survey", *Economic Development and Cultural Change,* **33** (1985), 255–298.

[139] Feder, G. and G. T. O'Mara, "Farm Size and the Diffusion of Green Revolution Technology", *Economic Development and Cultural Change,* **30 (1981), 59–76.

[140] Feder, G. and G. T. O'Mara, "On Information and Innovation Diffusion: A Bayesian Approach", *American Journal of Agricultural Economics,* **64 (1982), 145–147.

[141] Feder, G. and R. Slade, "The Acquisition of Information and the Adoption of New Technology", *American Journal of Agricultural Economics,* **66 (1984), 312–320.

[142] Feeny, D., "Technical and Institutional Change in Thai Agriculture, 1880–1940", *Ph.D. dissertation,* University of Wisconsin, Madison, 1976.

**[143] Feeny, D., *The Political Economy of Productivity: Thai Agricultural Development, 1880–1975.* Vancouver: University of British Columbia Press, 1981.

**[144] Feeny, D., *The Political Economy of Productivity: Thai Agricultural Development, 1880–1975.* Vancouver: University of British Columbia Press, 1982.

[145] Feeny, D., "Extensive versus Intensive Agricultural Development: Induced Public Investment in Southeast Asia, 1900–1940", *Journal of Economic History,* **43 (1983), 687–704.

[146] Feller, E. and D. C. Monzel, "Diffusion Millieus as a Focus of Research of Innovation in the Public Sector", *Policy Sciences,* **8 (1977), 49–68).

[147] Feller, I., *Approaches to the Diffusion of Innovations, Explanations in Entrepreneurial History,* Second Series, Vol. 4, No. 3, 1967.
[148] Fellner, W., *Trends and Cycles in Economic Activity.* New York: Holt, 1956.
[149] Fellner, W., "Two Propositions in the Theory of Induced Innovations", *Economic Journal,* **71** (1961), 305–308.
[150] Fellner, W., "Appraisal of the Labour Saving Character of Innovations", in *The Theory of Capital,* ed. by F. Lutz and D. Hague. London: Macmillan & Co., 1961.
[151] Fellner, W., "Does the Market Direct the Relative Factor-Saving Effects of Technological Progress"? in *The Rate and Direction of Inventive Activity: Economic and Social Factors.* Princeton: Princeton University Press for the National Bureau of Economic Research, 1962.
[152] Fellner, W., "Profit Maximization, Utility Maximization and the Rate and Direction of Innovation", *American Economic Review, Papers and Proceedings,* **56** (1966), 24–32.
[153] Fellner, W., "Comment on the Induced Bias", *Economic Journal,* **77** (1967), 662–664.
[154] Fellner, W., "Measures of Technological Progress in the Light of Recent Growth Theories", *American Economic Review,* **57** (1967), 1073–1098.
[155] Fellner, W., "Empirical Support for the Theory of Induced Innovations", *Quarterly Journal of Economics,* **85** (1971), 580–604.
[156] Ferguson, C. E., *The Neoclassical Theory of Production and Distribution.* London: Cambridge University Press, 1969.
[157] Field, A. J., "The Problem with Neoclassical Institutional Economics: A Critique with Special Reference to the North-Thomas Model of Pre 1500 Europe", *Explorations in Economic History,* **18** (1981), 174–198.
[158] Fischer, A. J. and R. K. Lindner, "The Effect of Distance to Information Source on Information Quality and the Time to Adoption", mimeographed, University of Adelaide, Adelaide, South Australia, 1980.
[159] Fisher, J. C. and R. H. Pry, "A Simple Substitution Model for Technological Change", *Technological Forecasting and Social Change,* **2** (1971), 75–88.
[160] Fixler, D. and U. Ben-Zion, "A Generalized Model of Induced Input Innovation", *European Economic Review,* **19** (1982), 277–287.
[161] Forsund, F. R. and E. S. Jansen, "Technical Progress and Structural Change in the Norwegian Primary Aluminum Industry", in *Topics in Production Theory,* ed. by F. R. Forsund. London: Macmillan, 1984.
[162] Frankel, F., *India's Green Revolution.* Princeton: Princeton University Press, 1971.
[163] Fransman, M., "Conceptualizing Technical Change in the Third World in the 1980s: An Interpretive Survey", *Journal of Development Studies,* **21** (1985), 572–652.
[164] Fundenberg, D. and J. Tirole, "Preemption and Rent Equalisation in the Adoption of New Technology", *Review of Economic Studies,* **52** (1985), 383–401.
[165] Gehrig, W., "On Certain Concepts of Neutral Technical Progress: Definitions, Implications, and Compatibility", Chapter 1 in *The Economics of Technological Progress,* ed. by T. Puu and S. Wibe. New York: St. Martin's Press, 1980.
[166] Gerhart, J., *The Diffusion of Hybrid Maize in West Kenya.* Mexico City: Centro Internacional de Mejoramiento de Maiz y Trigo, 1975.

**[167] Ghatak, S., "Technology Transfer to Developing Countries: The Case of the Fertilizer Industry", *Contemporary Studies in Economic and Financial Analysis,* Vol. 27. Greenwich, Conn.: JAI Press, 1981.
[168] Gibbons, M. and R. Johnston, "The Roles of Science in Technological Innovation", *Research Policy,* **3 (1974), 220–242.
**[169] Gilfillan, S. C., *Inventing the Ship.* Chicago: Follett, 1935.
**[170] Gilfillan, S. C., *The Sociology of Invention.* Chicago: Follett, 1935.
**[171] Gilfillan, S. C., "The Prediction of Inventions in Technological Trends and National Policy", National Resources Committee. Washington, D.C.: U.S. Government Printing Office, June 1937.
[172] Glaister, S., "Advertising Policy and Returns to Scale in Markets Where Information is Passed between Individuals", *Economica,* **41 (1974), 139–156.
[173] Globerman, S., "Technological Diffusion in the Canadian Tool and Die Industry", *Review of Economics and Statistics,* **57 (1975), 428–434.
[174] Godden, D., "Plant Variety Rights in Australia: Some Economic Issues", *Review of Marketing and Agricultural Economics,* **50 (1982), 51–59.
[175] Godden, D., "Plant Breeders' Rights and International Agricultural Research", *Food Policy,* **19 (1984), 206–218.
**[176] Godden, D., "Technological Change and Demand for Output at the Farm Level in U.K. Agriculture 1950–80", *Ph.D. dissertation,* London School of Economics, 1985.
[177] Gold, B., W. S. Pierce and G. Rosegger, "Diffusion of Major Technological Innovations in U.S. Iron and Steel Manufacturing", *Journal of Industrial Economics,* **18 (1970), 218–242.
[178] Gollup, F. M. and M. J. Roberts, "Environmental Regulations and Productivity Growth" The Case of Fossil-Fueled Electric Power Generation", *Journal of Political Economy,* **91 (1983), 654–674.
[179] Gomulka, S., "Do New Factories Embody Best Practice Technology? New Evidence", *Economic Journal,* **86 (1976), 859–863.
[180] Goodwin, R. M., "A Growth Cycle", in *Socialism, Capital and Economic Growth,* ed. by C. H. Feinstein. London: Cambridge University Press, 1967.
[181] Gort, M. and S. Klepper, "Time Paths in the Diffusion of Product Innovations", *Economic Journal,* **92 (1982), 630–653.
[182] Gort, M. and A. Konakayama, "A Model of Diffusion in the Production of an Innovation", *American Economic Review,* **72 (1982), 1111–1120.
**[183] Gould, J. P., "Diffusion Processes and Optimal Advertising Policy", in *Microeconomic Foundations of Employment and Inflation Theory,* by E. S. Phelps *et al.* New York: W. W. Norton, 1970.
[184] Grabowski, R., "The Implications of an Induced Innovation Model", *Economic Development and Cultural Change,* **27 (1979), 723–734.
**[185] Greene, B. A., *Rate of Adoption of New Farm Practices in the Central Plains, Thailand,* Cornell International Agricultural Bulletin No. 24, Cornell University, 1973, pp. 1–96.
[186] Greene, W. H., "Simultaneous Estimation of Factor Substitution, Economies of Scale, Productivity and non-Neutral Technical Change", in *Developments in Econometric Analysis of Productivity; Measurement and Modelling Issues,* ed. by A. Dogramaci. The Hague: Kluwer Nijhoff, 1983.
[187] Gregory, R. G. and D. W. James, "Do New Factories Embody Best Practice Technology?", *Economic Journal,* **83 (1973), 1133–1155.

**[188] Griffin, K., *The Political Economy of Agrarian Change,* 2nd edn. London: Macmillan & Co., 1979.

**[189] Grigg, D., *The Dynamics of Agricultural Change.* London: Hutchinson, 1982.

*[190] Griliches, Z., "Hybrid Corn: An Exploration in the Economics of Technological Change", *Econometrica,* **25** (1957), 501–522.

[191] Griliches, Z., "Research Costs and Social Returns: Hybrid Corn and Related Innovations", *Journal of Political Economy,* **66 (1958), 419–431.

[192] Griliches, Z., "Congruence Versus Profitability: A False Dichotomy", *Rural Sociology,* **25 (1960), 354–356.

[193] Griliches, Z., "Profitability Versus Interaction: Another False Dichotomy", *Rural Sociology,* **27 (1962), 327–330.

**[194] Griliches, Z., "Agriculture: Productivity and Technology", *International Encyclopedia of the Social Sciences,* Vol. 1. New York: Macmillan Co. and Free Press, 1968.

[195] Griliches, Z., "Hybrid Corn Revisited: A Reply", *Econometrica,* **48 (1980), 1463–1465.

[196] Gross, N., "The Differential Characteristics of Acceptance and Nonacceptance of an Approved Agricultural Technological Practice", *Rural Sociology,* **14 (1949), 148–156.

[197] Gruber, W. H., D. Mehta and R. Vernon: "The R & D Factor in International Trade and International Investment of U.S. Industries", *Journal of Political Economy,* **75** (1967), 20–37.

[198] Gruber, W. H. and R. Vernon, "The Technology Factor in a World Trade Matrix", in *The Technological Factor in International Trade,* ed. by R. Vernon pp. 233–272. New York: Columbia University Press for National Bureau of Economic Research, 1970.

[199] Gupta, K. S. and M. A. Taher, "Technological Progress and Factor Substitution in the U.S. Textile Industry 1949–1974", *Empirical Economics,* **9 (1984), 67–74.

[200] Gutkind, E. and D. Zilberman, "An Economic Approach to the Diffusion Process", *Indian Journal of Economics,* **65 (1985), 499–512.

[201] Guttman, Joel, "Interest Groups and the Demand for Agricultural Research", *Journal of Political Economy,* **86 (1978), 467–484.

[202] Guttman, J., "Villages as Interest Groups: The Demand for Agricultural Extension Services in India", *Kyklos,* **33 (1980), 122–141.

**[203] Habakkuk, H., *American and British Technology in the Nineteenth Century.* Cambridge: Cambridge University Press, 1962.

*[204] Hacche, G., *The Theory of Economic Growth: An Introduction.* London: Macmillan & Co., 1979.

**[205] Hagerstrand, T., *Innovation Diffusion as a Spatial Process.* Chicago: University of Chicago Press, 1967.

[206] Hahn, F. H. and R. C. O. Matthews, "The Theory of Economic Growth: A Survey", *Economic Journal,* **74 (1964), 779–902. Reprinted in American Economic Association/Royal Economic Society, *Surveys of Economic Theory,* Vol. 2. London: Macmillan & Co., 1965.

**[207] Hamilton, C. and H. T. Soderstrom, "Technology and International Trade: A Heckscher–Ohlin Approach", in *The World Economic Order: Past and Prospects,* ed. by S. Grassman and E. Lundberg. London: Macmillan & Co., 1981.

[208] Harcourt, G., *Some Cambridge Controversies in the Theory of Capital.* Cambridge: Cambridge University Press, 1972.

[209] Harley, C. K., "On the Persistence of Old Techniques: The Case of North American Wooden Shipbuilding", *Journal of Economic History,* **33 (1973), 372–398.
[210] Harrod, R., *Towards a Dynamic Economics.* London: Macmillan & Co., 1954.
[211] Hastings, T., "The Characteristics of Early Adopters of New Technology: An Australian Study", *Economic Record,* **52 (1976), 239–250.
[212] Havens, A. E. and E. M. Rogers, "Adoption of Hybrid corn: Profitability and the Interaction Effect", *Rural Sociology,* **26 (1961), 409–414.
[213] Hayami, Y., "Elements of Induced Innovations: A Historical Perspective for the Green Revolution", *Explorations in Economic History,* **8** (1971), 445–472.
[214] Hayami, Y., "Conditions for the Diffusion of Agricultural Technology: An Asian Perspective", *Journal of Economic History,* **34 (1974), 131–148.
[215] Hayami, Y., "Induced Innovation, Green Revolution and Income Distribution: Comment", *Economic Development and Cultural Change,* **30** (1981), 169–176.
[216] Hayami, Y. and R. Herdt, "Market Price Effects of Technological Change on Income Distribution in Semisubsistence Agriculture", *American Journal of Agricultural Economics,* **59** (1977), 245–256.
**[217] Hayami, Y. and M. Kikuchi, *Asian Village Economy at the Crossroads.* Tokyo: University of Tokyo Press, 1981.
[218] Hayami, Y. and V. Ruttan, "Factor Prices and Technical Change in Agricultural Development: The United States and Japan, 1880–1960", *Journal of Political Economy,* **78 (1970), 1115–1141.
**[219] Hayami, Y. and V. Ruttan, *Agricultural Development: An International Perspective.* Baltimore: Johns Hopkins University Press, 1971.
[220] Hayami, Y. and V. Ruttan, "Strategies for Agricultural Development", *Food Research Institute Studies in Agricultural Economics, Trade and Development,* **11 (1972), 129–148.
[221] Hayami, Y. and V. Ruttan, "Professor Rosenberg and the Direction of Technological Change: A Comment", *Economic Development and Cultural Change,* **21 (1973), 352–355.
*[222] Hayami, Y. and V. Ruttan, *Agricultural Development: An International Perspective,* rev. ed. Baltimore: Johns Hopkins University Press, 1985.
**[223] Hazell, P. B. and J. R. Anderson, "Public Policy Towards Technical Change in Agriculture", Chapter 8 in *Technology, Innovation and Economic Policy,* ed. by P. Hall. London: Philip Allen, 1986.
[224] Hernes, G., "Diffusion and Growth—the Non-Homogenous Case", *Scandinavian Journal of Economics,* **78 (1976), 427–436.
**[225] Hicks, J., *The Theory of Wages,* 1st ed. London: Macmillan & Co., 1932.
[226] Hicks, J., "Distribution and Economic Progress: A Revised Version", *Review of Economic Studies,* **4** (1936), 1–12.
[227] Hicks, J., *The Theory of Wages,* 2nd edn. London: Macmillan & Co., 1963.
**[228] Hicks, J., *Economic Perspectives: Further Essays in Money and Growth.* Oxford: Clarendon Press, 1977.
[229] Hiebert, L. D., "Risk, Learning and the Adoption of Fertilizer Responsive Seed Varieties", *American Journal of Agricultural Economics,* **56 (1974), 764–768.
[230] Hill, L. D. and P. Kau, "Application of Multivariate Probit to a Threshold Model of Grain Dryer Purchasing Decisions", *American Journal of Agricultural Economics,* **55 (1973), 19–25.

[231] Hirshleifer, J., "The Expanding Domain of Economics", *American Economic Review,* **75** (1985), 53–68.

[232] Horsky, D., "Comments on the Relationship between Diffusion Rates, Experience Curves and Demand Elasticities for Consumer Durable Technology Innovations", *Journal of Business,* Part II, **53 (1980), S75–S78.

**[233] Horsky, D. and L. S. Simon, "Advertising in a Model of New Product Diffusion", paper presented at the TIMS/ORSA National Meeting, New York City, May 1978.

**[234] Hughes, T. P., "Introduction", in *Selections from Lives of the Engineers,* ed. by S. Smiles. Cambridge, Mass.: MIT Press, 1966.

[235] Hunt, T. L., "The Structure and Changes of Technology in Prewar Japanese Agriculture: Comment", *American Journal of Agricultural Economics,* **62 (1980), 826–827.

**[236] Illinois Institute of Technology, "Technology in Retrospect and Critical Events in Science (TRACES)", mimeographed, Illinois Institute of Technology, Chicago, 1968.

[237] International Rice Research Institute, *Constraints to High Yields of Asian Rice Farms: An Interim Report,* Chapter 1. Los Banos, Philippines: International Rice Research Institute, 1978.

[238] Iwai, K., "Schumpeterian Dynamics, Part I: An Evolutionary Model of Innovation and Imitation", *Journal of Economic Behaviour and Organization,* **5 (1984), 159–190.

[239] Iwai, K., "Schumpeterian Dynamics, Part II: Technological Progress, Firm Growth and 'Economic Selection' ", *Journal of Economic Behaviour and Organization,* **5 (1984), 321–351.

**[240] Jamison, D. T. and L. J. Lau, *Farm Education and Farm Efficiency.* Baltimore: Johns Hopkins University Press, 1982.

[241] Jarvis, L. S., "Predicting the Diffusion of Improved Pastures in Uruguay", *American Journal of Agricultural Economics,* **63 (1981), 495–502.

[242] Jensen, R., "Adoption and Diffusion of an Innovation of Uncertain Profitability", *Journal of Economic Theory,* **27 (1982), 182–193.

[243] Jensen, R., "Innovation Adoption and Diffusion When There Are Competing Innovations", *Journal of Economic Theory,* **29 (1983), 161–171.

[244] Johnson, R., "Efficiency Growth in New Zealand Agriculture: A Review", *Economic Record,* **48 (1972), 76–91.

[245] Johnston, B. and J. Cownie, "The Seed-Fertilizer Revolution and Labor Force Absorption", *American Economic Review,* **59** (1969), 569–582.

[246] Jones, G. E., "The Adoption and Diffusion of Agricultural Practices", *World Agricultural Economics and Rural Sociology Abstract,* **9 (1967), 34.

**[247] Jones, H. G., *An Introduction to Modern Theories of Economic Growth.* New York: McGraw-Hill, 1976.

[248] Jones, R. W., " 'Neutral' Technical Change and the Isoquant Map", *American Economic Review,* **55 (1965), 848–855.

[249] Jones, R. W., "Comments on Technical Progress", *Philippine Economic Journal,* **5 (1966), 313–332.

**[250] Jones, R. W., "The Role of Technology in the Theory of International Trade", in *The Technology Factor in International Trade,* ed. by R. Vernon, pp. 73–92. New York: Columbia University Press for the National Bureau of Economic Research, 1970.

**[251] Jones, R. W., "A Three-Factor Model in Theory, Trade, and History", in *Trade, Balance of Payments and Growth,* ed. by J. Bhagwati, R. Mundell, and J. Vanek. Amsterdam: North-Holland, 1971.

**[252] Jorgenson, D. W. and B. M. Fraumeni, "Relative Prices and Technical Change", Chapter 2 in *Modelling and Measuring Natural Resource Substitution,* ed. by E. R. Berndt and B. C. Field. Cambridge, Mass.: MIT Press, 1981.

[253] Just, R. E. and R. Pope, "Stochastic Specification of of Production Functions and Economic Implications", *Journal of Econometrics,* **7 (1978), 67–86.

[254] Just, R. E. and D. E. Zilberman, "Stochastic Structure, Farm Size, and Technology Adoption in Developing Agriculture", *Oxford Economic Papers,* **35 (1983), 307–328.

**[255] Just, R. E., D. E. Zilberman and G. C. Rausser, "A Putty Clay Approach to the Distributional Effects of New Technology under Risk", in *Operations Research in Agriculture and Water Resources,* ed. by D. Yaron and C. Tapiero. Amsterdam: North-Holland, 1980.

[256] Kako, T., "Decomposition Analysis of Derived Demand for Factor Inputs: The Case of Rice Production in Japan", *American Journal of Agricultural Economics,* **60 (1978), 628–635.

[257] Kaldor, N., "A Model of Economic Growth", *Economic Journal,* **67 (1957), 591–624.

**[258] Kaldor, N., "Capital Accumulation and Economic Growth", in *The Theory of Capital,* ed. by F. A. Lutz and D. C. Hague. London: IEA, Macmillan & Co., 1961.

[259] Kaldor, N. and J. A. Mirrlees, "A New Model of Economic Growth", *Review of Economic Studies,* **29** (1962), 174–192.

[260] Kamien, M. and N. Schwartz, "Optimal Induced Technical Change", *Econometrica,* **36 (1968), 1–17.

[261] Kamien, M. and N. Schwartz, "Induced Factor Augmenting Technical Progress from a Microeconomic Viewpoint", *Econometrica,* **37 (1969), 668–684.

[262] Kamien, M. and N. Schwartz, "The Theory of the Firm with Induced Technical Change", *Metroeconomica,* **23 (1971), 233–256.

**[263] Kamien, M. and N. Schwartz, *Market Structure and Innovation.* Cambridge: Cambridge University Press, 1982.

[264] Kaneda, H., "Specification of Production Functions for Analysing Technical Change and Factor Inputs in Agricultural Development", *Journal of Development Economics,* **11** (1982), 97–108.

[265] Katz, E., H. Hamilton and M. L. Levin, "Traditions of Research on the Diffusion of Innovation", *American Sociological Review,* **28 (1963), 237–252.

[266] Kawagoe, T., K. Otsuka and Y. Hayami, "Induced Biases of Technical Change in Agriculture: The U.S. and Japan, 1880–1980", *Journal of Political Economy,* **94 (1986) 523–544.

*[267] Kelly, P. and M. Kransberg, eds, *Technological Innovation: A Critical Review of Current Knowledge.* San Francisco: San Francisco Press, 1978.

**[268] Kendrick, J. W. and B. N. Vaccara, eds, *New Developments in Productivity Measurement and Analysis.* Chicago: University of Chicago Press, 1980.

[269] Kennedy, C., "Induced Bias in Innovation and the Theory of Distribution", *Economic Journal,* **74 (1964), 541–547.

[270] Kennedy, C., "Samuelson on Induced Innovation", *Review of Economics and Statistics,* **48** (1966), 442–444.

[271] Kennedy, C., "A Generalisation of the Theory of Induced Bias in Technical Progress", *Economic Journal,* **83 (1973), 48–57.

*[272] Kennedy, C. and A. Thirlwall, "Technical Progress: A Survey", *Economic Journal,* **82** (1972), 11–72.
**[273] Kenwood, A. G. and A. L. Lougheed, *Technological Diffusion and Industrialisation before 1914.* London: Croom Helm, 1982.
[274] King, R. E., "The Frontier Production Function: A Tool for Improved Decision Making", *Journal of the Northeastern Agricultural Economics Council,* **9** (1980), 1–10.
[275] Kislev, Y. and W. Peterson, "Induced Innovation and Farm Mechanisations", *American Journal of Agricultural Economics,* **63** (1981), 562–565.
[276] Kislev, Y. and W. Peterson, "Prices, Technology and Farm Size", *Journal of Political Economy,* **90 (1982), 578–595.
[277] Kislev, Y. and N. Schori-Bachrach, "The Process of an Innovation Cycle", *American Journal of Agricultural Economics,* **55 (1973), 28–37.
[278] Klein, K. K. and E. W. Kehrberg, "The Use of an Innovation Possibility Frontier to Evaluate an Applied Animal Breeding Research Project", *Canadian Journal of Agricultural Economics,* **29 (1981), 141–158.
[279] Klingaman, D., R. Vedder and L. Gallaway, "The Ames-Rosenberg Hypothesis Revisited", *Explorations in Economic History,* **11 (1974), 311–314.
**[280] Kuhn, T. S., *The Structure of Scientific Revolutions.* Chicago: University of Chicago Press, 1962.
**[281] Kuznets, S., *Secular Movements in Production and Prices.* Boston: Houghton Mifflin, 1930.
**[282] Kuznets, S., *Econimic Change.* London: Heinemann 1954.
**[283] Kuznets, S., "Technological Innovations and Economic Growth", Chapter 14 in *Technological Innovation: A Critical Review of Current Knowledge,* ed. by P. Kelly and M. Kransberg, pp. 335–362. San Francisco: San Francisco Press, 1978.
[284] Ladejinsky, W., "The Green Revolution in Punjab: A Field Trip", *Economic and Political Weekly,* **4** (1969), A73–A82.
[285] Laibman, D., "Two-Sector Growth with Endogenous Technical Change: A Marxian Analysis", *Quarterly Journal of Economics,* **96** (1981), 45–75.
[286] Lau, J. L. and S. Tamura, "Economies of Scale, Technical Progress, and the Nonhomothetic Leontief Production Function: An Application to the Japanese Petrochemical Processing Industry", *Journal of Political Economy,* **80 (1972), 1167–1187.
[287] Layton, E. T. Jr, "Scientific Technology, 1845–1900: The Hydraulic Turbine and the Origins of American Industrial Research", *Technology and Culture,* **20 (1979), 64–89.
**[288] Lazarsfeld, P. F., B. Berelson and H. Gaudet, *The People's Choice.* New York: Columbia University Press, 1948.
[289] Lazonick, W., "Factor Costs and the Diffusion of Ring Spinning in Britain Prior to World War I", *Quarterly Journal of Economics,* **96 (1981), 89–109.
[290] Lee, J., "The Measurement and Sources of Technological Change Biases with an Application to Postwar Japanese Agriculture", *Economica,* **50 (1983), 159–174.
**[291] Leik, R. K. and B. F. Meeker, *Mathematical Sociology.* Englewood Cliffs, N. J.: Prentice-Hall, 1975.
[292] Lekvall, P. and C. Wahlbin, "A Study of Some Assumptions Underlying Innovation Diffusion Function", *Swedish Journal of Economics,* **75 (1973), 362–377.
**[293] Levy, V., "Total Factor Productivity, Non-Neutral Technical Change and

Economic Growth: A Parametric Study of a Developing Economy", *Journal of Development Economics,* **8** (1981), 93–109.
[294] Lianos, T. P., "The Relative Shares of Labor in U.S. Agriculture", *American Journal of Agricultural Economics,* **53 (1971), 411–422.
[295] Lilien, G., "The Implications of Diffusion Models for Accelerating the Diffusion of Innovation", *Technological Forecasting and Social Change,* **17 (1980), 339–351.
**[296] Lindner, R. K., "Farm Size and the Time Lag to Adoption of a Scale Neutral Innovation", mimeographed, University of Adelaide, Adelaide, South Australia, 1980.
[297] Lindner, R. K., A. Fischer and P. Pardey, "The Time to Adoption", *Economic Letters,* **2 (1979), 187–190.
[298] Lipton, M., "The Theory of the Optimising Peasant", *Journal of Development Studies,* **4** (1968), 327–351.
[299] Lopez, R. E., "The Structure of Production and the Derived Demand for Inputs in Canadian Agriculture", *American Journal of Agricultural Economics,* **62 (1980), 38–45.
[300] Lucas, R. E., "Tests of a Capital–Theoretical Model of Technical Change", *Review of Economic Studies,* **34** (1967), 175–190.
[301] Lynk, E., "Factor Demand, Substitution and Biased Technical Change in Indian Manufacturing Industries", *Manchester School,* **50 (1982), 126–138.
**[302] Maclaurin, W. R., *Invention and Innovation in the Radio Industry.* New York: Macmillan Co., 1949.
[303] Maclaurin, W. R., "Technological Progress in Some American Industries", *American Economic Review,* **44** (1954), 178–200.
**[304] Maddala, G. S., *Econometrics.* New York: McGraw-Hill, 1977.
**[305] Maddala, G. S., *Limited-Dependent and Qualitative Variables in Econometrics.* London: Cambridge University Press, 1983.
[306] Magat, W., "Regulation and the Rate and Direction of Induced Technical Change", *Bell Journal of Economics,* **7 (1976), 478–496.
[307] Magat, W., "Technological Advance with Depletion of Innovation Possibilities—Implications for the Dynamics of Factor Shares", *Economic Journal,* **89 (1979), 615–623.
[308] Mahajan, V. and R. A. Peterson, "Innovation Diffusion in a Dynamic Potential Adopter Population", *Management Science,* **24 (1978), 1589–1593.
[309] Mahajan, V. and R. A. Peterson, "Integrating Time and Space in Technological Substitution Models", *Technological Forecasting and Social Change,* **14 (1) (1979), 231–241.
[310] Mahajan, V. and R. A. Peterson, "First Purchase Diffusion Models of New-Product Acceptance", *Technological Forecasting and Social Change,* **15 (2) (October 1979), 127–146.
*[311] Mahajan, V. and R. A. Peterson, *Models for Innovation Diffusion.* Beverly Hills, Calif.: Sage Publications, 1985.
[312] Mahajan, V. and M. E. F. Schoeman, "Generalized Model for the Time Pattern of the Diffusion Process", *IEEE (Institute of Electrical and Electronics Engineers) Transactions of Engineering Management,* **24 (1977), 12–18.
[313] Mansfield, E., "Technical Change and the Rate of Imitation", *Econometrica,* **2 (1961), 741–766.
[314] Mansfield, E., "Intrafirm Rates of Diffusion of an Innovation", *Review of Economics and Statistics,* **45 (1963), 348–359.

[315] Mansfield, E., "The Speed of Response of Firms to New Techniques", *Quarterly Journal of Economics,* **77** (1963), 290–311.
[316] Mansfield, E., *Industrial Research and Technological Innovation.* New York: Norton, 1968.
[317] Mansfield, E., "Determinants of the Speed of Application of New Technology", Chapter 8 in *Science and Technology in Economic Growth,* ed. by Williams. New York: Wiley, 1973.
[318] Mansfield, E., J. Rapaport, J. Schnee, S. Wagner and M. Hamburger, *Research and Innovation in the Modern Corporation.* New York: Norton, 1971.
[319] Mansfield, E., J. Rapaport, A. Romeo, E. Villani, S. Wagner and R. Husic, *The Production and Application of New Industrial Technology.* New York: Norton, 1977.
[320] Mantoux, P., *The Industrial Revolution in the Eighteenth Century,* rev. ed. London: Jonathan Cape, 1928.
[321] Mark, J., G. M. Chapman and J. Gibson, "Bioeconomics and the Theory of Niches", *Futures* (December 1985), pp. 632–651.
[322] Marshall, Alfred, *Principles of Economics,* 8th edn. London: Macmillan & Co., 1948 (first published, 1890).
[323] McCain, R. A., "Land in Fellner's Neoclassical Model of Economic Growth: Comment", *American Economic Review,* **60** (1970), 495–499.
[324] McCain, R. A., "Induced Technical Progress and the Price of Capital Goods", *Economic Journal,* **82** (1972), 921–933.
[325] McCain, R. A., "Induced Bias in Technical Innovation Including Product Innovation in a Model of Economic Growth", *Economic Journal,* **84** (1974), 959–966.
[326] McCain, R. A., "The Characteristics of Optimum Inventions: An Isotech Approach", *American Economic Review,* **67** (1977), 365–369.
[327] McCain, R. A. "Endogenous Bias in Technical Progress and Environment Policy", *American Economic Review,* **68** (1978), 538–546.
[328] McCloskey, D. N., ed., *Essays on a Mature Economy: Britain after 1840.* Princeton, N.J.: Princeton University Press, 1971.
[329] McKay, L., D. Lawrence, and C. Vlastuin, "Production Flexibility and Technical Change in Australia's Wheat-Sheep Zone", *Review of Marketing and Agricultural Economics,* **50** (1982), 9–26.
[330] McLean, I., "The Analysis of Agricultural Productivity: Alternative Views and Victorian Evidence", Working Paper, Economics Department, University of Adelaide, Adelaide, South Australia, 1979.
[331] Metcalfe, J. S., "Diffusion of Innovation in the Lancashire Textile Industry", *Manchester School,* **38** (1970), 145–162.
[332] Metcalfe, J. S., "Impulse and Diffusion in the Study of Technical Change", *Futures,* **13** (1981), 347–359.
[333] Metcalfe, J. S., "On the Diffusion of Innovation and the Evolution of Technology", mimeographed, University of Manchester, 1982.
[334] Metcalfe, J. S., "On Technological Competition", paper presented to International Colloquim on Structural Change and New Technology, University of Manchester, September 10–13, 1985.
[335] Metcalfe, J. S., "Technological Innovation and the Competitive Process", Chapter 2 in *Technology, Innovation and Economic Policy,* ed. by P. Hall. London: Philip Allen, 1986.
[336] Metcalfe, J. S. and M. Gibbons, "On the Economics of Structural Change

and the Evolution of Technology", paper presented at Seventh World Congress of the International Economics Association, Madrid, September 1983.

**[337] Metcalfe, J. S. and L. Soete, "Notes on the Evolution of Technology and International Competition", Chapter 15 in *Science and Technology Policy in the 1980s and Beyond,* ed. by M. Gibbons, P. Gummett, and B. M. Udgaonkar. London; New York: Longman, 1984.

**[338] Miller, R. and D. Sawers, *The Technical Development of Modern Aviation.* London: Routledge & Kigan Paul, 1968.

**[339] Mishan, E. J., *The Costs of Economic Growth.* New York: Praeger, 1967.

**[340] Mohr, L., *Explaining Organizational Behavior.* San Francisco: Jossey-Bass, 1982.

[341] Mokyr, J., "Demand vs. Supply in the Industrial Revolution" *Journal of Economic History,* **37 (1977), 981–1008.

[342] Mokyr, J., "Technological Change and Economic Growth", in *The Economics of Technological Change,* ed. by F. M. Scherer, 1986.

**[343] Mooney, P. R., "The Law of the Seed: Another Development and Plant Genetic Resources", *Development Dialogue,* No. 1–2 (1983), 1–172.

**[344] Morishima, M. and M. Saito, "An Economic Test of Sir John Hicks' Theory of Biased Induced Investigations", in *Value, Capital and Growth: Papers in Honour of Sir John Hicks,* ed. by J. Wolfe. Chicago: Aldine Publishing Co., 1968.

**[345] Moroney, J. R. and J. M. Trapani, "Alternative Models of Substitution and Technical Change in Natural Resource Intensive Industries", Chapter 3 in *Modelling and Measuring Natural Resource Substitution,* ed. by E. R. Berndt and B. C. Field. Cambridge, Mass.: MIT Press, 1981.

[346] Mowery, D. C. and N. Rosenberg, "The Influence of Market Demand upon Innovation: A Critical Review of Some Recent Empirical Studies", *Research Policy,* **8 (1979), 103–153. Reprinted in N. Rosenberg, *Inside the Black Box: Technology and Economics,* Chapter 10, pp. 193–241. New York: Cambridge University Press, 1982.

[347] Mueller, C., "Factor Prices and Labor-Saving Technology in Brazil's Agriculture", *Journal of Economic Studies,* **7 (1980), 109–121.

**[348] Mundell, R. A., *International Economics.* New York: Macmillan Co., 1968.

[349] Nadiri, M. I., "Some Approaches to the Theory and Measurement of Total Factor Productivity", *Journal of Economic Literature,* **8 (1970), 1137–1177.

**[350] Nasbeth, L. and G. F. Ray, *Diffusion of New Industrial Processes.* Cambridge: Cambridge University Press, 1974.

[351] Nelson, R., "The Economics of Invention: A Survey of the Literature", *Journal of Business,* **32 (1959), 101–127.

[352] Nelson, R., "A Diffusion Model of International Productivity Differences in Manufacturing Industry", *American Economic Review,* **58** (1968), 1219–1247.

[353] Nelson, R., "Production Sets, Technological Knowledge, and R & D: Fragile and Overworked Constructs for Analysis of Productivity Growth?", *American Economic Review, Papers and Proceedings,* **70 (1980), 82–71.

[354] Nelson, R., "Research on Productivity Growth and Differences: Dead Ends and New Departures", *Journal of Economic Literature,* **19 (1981), 1029–1064.

**[355] Nelson, R. and E. S. Phelps, "Investment in Humans, Technological

Diffusion and Economic Growth", *American Economic Review,* **56** (1966), 69–75.
[356] Nelson, R. and S. Winter, "Toward an Evolutionary Theory of Economic Capabilities", *American Economic Review, Papers and Proceedings,* **63** (1973), 440–449.
[357] Nelson, R. and S. Winter, "Neoclassical vs. Evolutionary Theories of Economic Growth: Critique and Prospectus", *Economic Journal,* **84** (1974), 886–905.
[358] Nelson, R. and S. Winter, "In Search of a Useful Theory of Innovation", *Research Policy,* **6** (1977), 36–76.
[359] Nelson, R. and S. Winter, "Forces Generating and Limiting Concentration under Schumpeterian Competition", *Bell Journal of Economics,* **9** (1978), 524–548.
[360] Nelson, R. and S. Winter, *An Evolutionary Theory of Economic Change.* Cambridge, Mass.: Belknap Press of Harvard University, 1982.
[361] Nelson, R. and S. Winter, "The Schumpeterian Tradeoff Revisited", *American Economic Review,* **72** (1982), 114–132.
[362] Nelson, R., S. Winter and H. Schuette, "Technical Change in an Evolutionary Model", *Quarterly Journal of Economics,* **40** (1976), 90–118.
[363] Nerlove, M. and S. J. Press, "Multivariate Log-Linear Probability Models for the Analysis of Qualitative Data", mimeographed, Northwestern University, Evanston, Illinois, 1973.
[364] Nghiep, L., "The Structure and Changes in Technology in Prewar Japanese Agriculture", *American Journal of Agricultural Economics,* **61** (1979), 686–693.
[365] Nghiep, L., "The Structure and Changes of Technology in Prewar Japanese Agriculture: Reply", *American Journal of Agricultural Economics,* **62** (1980), 828–829.
[366] Nordhaus, W. D., "The Optimal Rate and Direction of Technical Change", in *Essays in the Theory of Optimal Economic Growth,* ed. by K. Shell. Cambridge, Mass.: MIT Press, 1967.
[367] Nordhaus, W. D., "The Optimal Rate and Direction of Technological Change", Chapter 6 in *Invention, Growth and Welfare* by W. D. Nordhaus. Cambridge, Mass.: MIT Press, 1969.
[368] Nordhaus, W. D., "Some Skeptical Thoughts on the Theory of Induced Innovation", *Quarterly Journal of Economics,* **87** (1973), 209–219.
[369] Norsworthy, J. R. and D. H. Malmquist, "Input Measurement and Productivity Growth in Japanese and U.S. Manufacturing", *American Economic Review,* **73** (1983), 947–967.
[370] North, D. C. and R. P. Thomas, "An Economic Theory of the Growth of the Western World", *Economic History Review,* **22** (1970), 1–17.
[371] North, D. C. and R. P. Thomas, *The Rise of the Western World.* Cambridge: Cambridge University Press, 1973.
[372] Norris, K. and J. Vaizey, *The Economics of Research and Technology.* London: George Allen and Unwin, 1973.
[373] Nove, A., *An Economic History of the USSR.* London: Penguin, 1969.
[374] Office of the Director of Defense and Engineering, *Project Hindsight: Final Report (HINDSIGHT).* Washington, D.C., 1969.
[375] Ogburn, W. F., *Social Change.* New York: Viking Press, 1922.
[376] Ogburn, W. F. and D. F. Nimoff, *Sociology.* New York: Houghton Mifflin, 1940.

[377] Ogburn, W. F. and D. S. Thomas, "Are Inventions Inevitable?", *Political Science Quarterly,* **37 (1922), 83–98.
[378] Okuguchi, K., "The Implications of Regulation for Induced Technical Change: Comment", *Bell Journal of Economics,* **6 (1975), 703–705.
[379] Oliver, F. R., "Methods of Estimating the Logistic Growth Function", *Applied Statistics,* **13 (1964), 57–66.
[380] Olmstead, A. L., "The Mechanization of Reaping and Mowing in American Agriculture, 1833–1870", *Journal of Economic History,* **35 (1975), 327–352.
[381] Olsen, J. and S. Choi, "A Product Diffusion Model Incorporating Repeat Purchases", *Technological Forecasting and Social Change,* **27 (1985), 385–397.
**[382] O'Mara, G., "A Decision–Theoretic View of Technique Diffusion in a Developing Country", *Ph.D. dissertation,* Stanford University, 1971.
[383] Orosel, G., "Capital Gains and Losses and the Existence of a Steady State in Multisectoral Models with Induced Technological Progress", *Economic Journal,* **87 (1977), 315–323.
[384] Oster, S., "The Diffusion of Innovation among Steel Firms: The Basic Oxygen Furnace", *Bell Journal of Economics,* **13 (1982), 45–56.
[385] Oster, S. M. and J. M. Quigley, "Regulatory Barriers to the Diffusion of Innovation: Some Evidence from Building Codes", *Bell Journal of Economics,* **8 (1977), 361–377.
[386] Ozga, S., "Imperfect Markets through Lack of Knowledge", *Quarterly Journal of Economics,* **74 (1960), 29–52.
[387] Panik, M. J., "Factor Learning and Biased Factor-Efficiency Growth in the United States, 1929–66", *International Economic Review,* **17 (1976), 733–739.
**[388] Park, J. K., "Technological Change in the Korean Rice Economy: Sources, Direction and Impact", *Ph.D. dissertation,* University of Minnesota, 1986.
[389] Parker, W. N., "Review of American and British Technology", *Business History Review,* **37** (1963), 121–122.
**[390] Pasinetti, L., *Structural Change and Economic Growth.* Cambridge: Cambridge University Press, 1981.
[391] Peterson, R. A. and V. Mahajan, "Multi-Product Growth Models", in *Research in Marketing,* ed. by J. Sheth. Greenwich, Conn.: JAI Press, 1978.
**[392] Peterson, W. and Y. Hayami, "Technical Change in Agriculture", in *A Survey of Agricultural Economics Literature,* Vol. 1, Part VII, ed. by L. R. Martin. Minneapolis: University of Minnesota Press, 1977.
[393] Phelps, E., "Models of Technical Progress and the Golden Rule of Research", *Review of Economic Studies,* **33 (1966), 133–145.
**[394] Phillips, W. H., "The Economic Performance of Late Victorian Britain: Traditional Historians and Growth", manuscript, 1981, (available from Phillips).
[395] Phillips, W. H., "Induced Innovation and Economic Performance in Late Victorian British Industry", *Journal of Economic History,* **42 (1982), 97–103.
**[396] Pindyck, R. S. and D. L. Rubinfeld, *Econometric Models and Economic Forecasts,* 2nd ed New York: McGraw-Hill, 1981.
[397] Piore, M., "The Impact of the Labour Market upon the Design and Selection of Productive Techniques within the Manufacturing Plant", *Quarterly Journal of Economics,* **82 (1968), 602–620.
**[398] Polanyi, K., *The Great Transformation.* Boston: Beacon Press, 1957.

[399] Posner, M. V., "International Trade and Technical Change", *Oxford Economic Papers,* **13** (1961), 323–341.
[400] Price, D. de S., "Is Technology Historically Independent of Science? A Study in Statistical Historiography", *Technology and Culture,* **6 (1965), 533–568.
**[401] Pugel, T. A., "Technology Transfer and the Neo-classical Theory of International Trade", in *Technology Transfer and Economic Development,* ed. by R. Hawkins and A. Prasad. Greenwich, Conn.: JAI Press, 1981.
**[402] Pyatt, F. G., *Priority Patterns and the Demand for Household Durable Goods.* Cambridge: Cambridge University Press, 1964.
[403] Pyle, J. W. and S. J. Turnovsky, "Risk Aversion in Chance Constrained Portfolio Selection", *Management Science,* **18 (1971), 218–225.
[404] Quizon, J. and H. Binswanger, "Income Distribution in Agriculture: A Unified Approach", *American Journal of Agricultural Economics,* **65** (1983), 526–538.
[405] Radner, R., "A Behavioral Model of Cost Reduction", *Bell Journal of Economics,* **6 (1975), 196–215.
[406] Rahm, M. R. and W. E. Huffman, "The Adoption of Reduced Tillage: The Role of Human Capital and Other Variables", *American Journal of Agriculture Economics,* **66 (1984), 405–413.
[407] Rao, P. S. and R. S. Preston, "Inter-Factor Substitution, Economies of Scale and Technical Change: Evidence from Canadian Industries", *Empirical Economics,* **9 (1984), 87–111.
[408] Rao, R. and D. Rutenberg, "Preempting an Alert Rival: Strategic Timing of the First Plant by Analysis of Sophisticated Rivalry", *Bell Journal of Economics,* **10 (1979), 412–428.
[409] Rapaport, J., "Diffusion of Technological Innovation among Non-Profit Firms: A Study of Radioisotopes in U.S. Hospitals", *Journal of Economics and Business,* **30 (1978), 108–118.
**[410] Ray, G. F., "The Diffusion of Mature Technologies", NIESR Occasional Paper No. 36, Cambridge University Press, 1984.
[411] Ray, S. C., "A Translog Cost Function Analysis of U.S. Agriculture, 1939–77", *American Journal of Agricultural Economics,* **64** (1982), 490–498.
[412] Reder, M., "Chicago Economics: Permanence and Change", *Journal of Economic Literature,* **20 (1982), 1–38.
[413] Reinganum, J. F., "Market Structure and the Diffusion of New Technology", *Bell Journal of Economics,* **12 (1981), 618–624.
[414] Reinganum, J. F., "On the Diffusion of New Technology: A Game Theoretic Approach", *Review of Economic Studies,* **48 (1981), 395–405.
[415] Reinganum, J. F., "Technology Adoption under Imperfect Information", *Bell Journal of Economics,* **14 (1983), 57–69.
[416] Resek, R. W., "Neutrality of Technical Progress", *Review of Economics and Statistics,* **45 (1963), 55–63.
[417] Robinson, J., "The Classification of Inventions", *Review of Economic Studies,* **5 (1938), 139–142.
**[418] Robinson, J., *The Accumulation of Capital.* London: Macmillan & Co., 1956.
[419] Robinson, V. and C. Lakhani, "Dynamic Price Models for New-Product Planning", *Management Science,* **21 (1975), 113–1122.
**[420] Rogers, E. M., *Diffusion of Innovations.* New York: Free Press, 1962.
[421] Rogers, E. M., *Modernization among Peasants: The Impact of Communications.* New York: Holt, Rinehart and Winston, 1969.

**[422] Rogers, E. M., "Barefoot Doctors", in *Rural Health in the People's Republic of China,* ed. by G. Lythscott. Washington, D.C.: U.S. Department of Health and Human Services, 1980.

*[423] Rogers, E. M., *Diffusion of Innovations,* 3rd ed. New York: Macmillan Co., 1983.

**[424] Rogers, E. M. and J. D. Eveland, "Diffusion of Innovations Perspective on National R & D Assessment: Communication and Innovation in Organizations", Chapter 12 in *Technological Innovation: A Critical Review of Current Knowledge,* ed. by P. Kelly and M. Kransberg. San Francisco: San Francisco Press, 1978.

[425] Rogers, E. M. and A. E. Havens, "Rejoinder to Griliches' 'Another False Dichotomy'", *Rural Sociology,* **27** (1962), 330–332.

[426] Romeo, A. A., "Interindustry and Interfirm Differences in the Rate of Diffusion of an Innovation", *Review of Economics and Statistics,* **57 (1975), 311–319.

[427] Romeo, A. A., "The Rate of Imitation of a Capital-Embodied Process Innovation", *Economica,* **44 (1977), 63–69.

**[428] Rosegger, G., *The Economics of Production and Innovation: An Industrial Perspective.* Elmsford, N.Y.: Pergamon Press, 1980.

[429] Rosenberg, N., "The Direction of Technological Change: Inducement Mechanisms and Focusing Devices", *Economic Development and Cultural Change,* **18 (1969), 1–24.

*[430] Rosenberg, N., ed., *The Economics of Technological Change.* Harmondsworth, Middlesex, England: Penguin, 1971.

[431] Rosenberg, N., "Science, Invention and Economic Growth", *Economic Journal,* **100 (1974), 725–729.

*[432] Rosenberg, N., *Perspective on Technology.* New York: Cambridge University Press, 1976.

**[433] Rosenberg, N., "Problems in the Economists' Conceptualization of Technological Innovation", Chapter 4 in *Perspectives on Technology.* Cambridge: Cambridge University Press, 1976.

*[434] Rosenberg, N., *Inside the Black Box: Technology and Economics.* Cambridge: Cambridge University Press, 1982.

**[435] Rosenberg, N., "The Historiography of Technical Progress", Chapter 1, in *Inside the Black Box: Technology and Economics,* ed. by N. Rosenberg, pp. 3–33. Cambridge: Cambridge University Press, 1982.

**[436] Rosenberg, N., "On Technological Expectations", Chapter 5, in *Inside the Black Box: Technology and Economics,* ed. by N. Rosenberg, pp. 104–119. Cambridge: Cambridge University Press, 1982.

**[437] Rosenberg, N., "Learning by Using", Chapter 6, in *Inside the Black Box: Technology and Economics,* ed. by N. Rosenberg, pp. 120–140. Cambridge: Cambridge University Press, 1982.

**[438] Rosenberg, N., "How Exogenous Is Science?", Chapter 7, in *Inside the Black Box: Technology and Economics,* ed. by N. Rosenberg, pp. 141–149. Cambridge: Cambridge University Press, 1982.

[439] Rothbarth, E., "Causes of the Superior Efficiency of U.S.A. Industry as Compared with British Industry", *Economic Journal,* **56 (1946), 383–390.

**[440] Roumasset, J., *Rice and Risk: Decision Making among Low Income Farmers.* Amsterdam: North-Holland, 1976.

[441] Russell, L. B., "The Diffusion of Hospital Technologies: Some Econometric Evidence", *Journal of Human Research,* **12 (1977), 482–502.

[442] Russell, L. B. and C. S. Burke, *Technological Diffusion in the Hospital Sector*. Chicago: National Planning Association, 1975.
[443] Russell, T., "Comments on: The Relationship between Diffusion Rates, Experience Curves and Demand Elasticities for Consumer Durable Technical Innovations", *Journal of Business,* **52** (1980), S69–S73.
[444] Ruttan, V. W., "The Contribution of Technological Progress to Farm Output: 1950–75", *Review of Economics and Statistics,* **38** (1956), 61–69.
[445] Ruttan, V. W., "Usher and Schumpeter on Invention, Innovation and Technological Change", *Quarterly Journal of Economics,* **73** (1959), 596–606.
[446] Ruttan, V. W., "Technology and the Environment", *American Journal of Agricultural Economics,* **53** (1971), 707–717.
[447] Ruttan, V. W., "Technology Transfer, Institutional Transfer, and Induced Technical and Institutional Change in Agricultural Development", Chapter 7 in *Agriculture in Development Theory,* ed. by L. G. Reynolds. New Haven: Yale University Press, 1975.
[448] Ruttan, V. W., "The Green Revolution: Seven Generalizations", *International Development Review,* **19** (1977), 16–23.
[449] Ruttan, V. W., "Induced Institutional Change", Change 12 in *Induced Innovation,* ed. by H. Binswanger and V. W. Ruttan. Baltimore: Johns Hopkins University Press, 1978.
[450] Ruttan, V. W., "Induced Institutional Innovation", *Agricultural Economics Research,* **31** (3) (1979), 32–35.
[451] Ruttan, V. W., "An Induced Innovation Interpretation of Technical Change in Agriculture in Developed Countries", Instituto Interamericano de Cooperacion para la Agricultura, Costa Rica, 1981.
[452] Ruttan, V. W., *Agricultural Research Policy*. Minneapolis: University of Minnesota Press, 1982.
[453] Ruttan, V. W., "Social Science Knowledge and Institutional Change", *American Journal of Agricultural Economics,* **66** (1984), 549–559.
[454] Ruttan, V. W. and H. Binswanger, "Induced Innovation and the Green Revolution", Chapter 13 in *Induced Innovation,* ed. by H. Binswanger and V. W. Ruttan. Baltimore: Johns Hopkins University Press, 1978.
[455] Ruttan, V. W., H. Binswanger, Y. Hayami, W. Wade and A. Weber, "Factor Productivity and Growth: A Historical Interpretation", Chapter 3 in *Induced Innovation,* ed. by H. Binswanger and V. W. Ruttan. Baltimore: Johns Hopkins University Press, 1978.
[456] Ruttan, V. W. and Y. Hayami, "Technology Transfer and Agricultural Developments", *Technology and Culture,* **14** (1973), 119–151.
[457] Ruttan, V. W. and Y. Hayami, "Towards a Theory of Induced Institutional Innovation", *Journal of Development Studies,* **20** (1984), 203–223.
[458] Ryan, B. and N. Gross, "The Diffusion of Hybrid Seed Corn in Two Iowa Communities", *Rural Sociology,* **8** (1943), 15–24.
[459] Sahal, D., *Patterns of Technological Innovation*. Reading, Mass.: Addison-Wesley Publishing Co., 1981.
[460] Sahota, B. W., Review of *Agricultural Development: An International Perspective,* ed. by Y. Hayami and V. W. Ruttan, *Economic Development and Cultural Change,* **21** (1973), 722–731.
[461] Salter, W. E. G., *Productivity and Technical Change,* 1st ed. Cambridge: Cambridge University Press, 1960.
[462] Salter, W. E. G., *Productivity and Technical Change,* 2nd ed. Cambridge: Cambridge University Press, 1966.

[463] Samuelson, P., "A Theory of Induced Innovation Along Kennedy–Weisacker Lines", *Review of Economics and Statistics,* **47 (1965), 343–356.

[464] Samuelson, P., "Rejoinder: Agreements, Disagreements, Doubts, and the Case of Induced Harrod-Neutral Technical Change", *Review of Economics and Statistics,* **48 (1966), 444–448.

**[465] Sandberg, L. G., *Lancashire in Decline: A Study in Entrepreneurship, Technology and International Trade.* Columbus: Ohio State University Press, 1974.

**[466] Sandberg, L. G., "The Entrepreneur and Technological Change", Chapter 5 in *The Economic History of Britain Since 1700, Vol. 2, 1860 to the 1970s,* ed. by R. Floud and D. McCloskey, 1981. New York: Cambridge University Press, 1981.

**[467] Sanders J. and V. Ruttan, "Biased Choice of Technology in Brazilian Agriculture", Chapter 10 in *Induced Innovation,* ed. by H. Binswanger and V. W. Ruttan. Baltimore: Johns Hopkins University Press, 1978.

[468] Sato, R., "The Estimation of Biased Technical Progress and the Production Function", *International Economic Review,* **11 (1970), 179–208.

[469] Sato, R., *Theory of Technical Change and Economic Invariance: Application of Lie Groups.* New York: Academic Press, 1981.

[470] Sato, R. and M. J. Beckman, "Neutral Innovations and Production Functions", *Review of Economic Studies,* **35** (1968), 57–66.

**[471] Sato, R. and G. S. Suzawa, *Research and Productivity: Endogenous Change.* Boston: Auburn House, 1983.

**[472] Sauer, C. O., *Agricultural Origins and Dispersals: The Domestication of Animals and Foodstuffs,* 2nd edn. Cambridge, Mass.: MIT Press, 1969.

[473] Saul, S., "The Market and the Development of the Mechanical Engineering Industries in Britain, 1860–1914", *Economic History Review,* 2nd series, **20** (1967), 111–130.

**[474] Saul, S., ed., Editor's Introduction in *Technological Change, the United States and Britain in the Nineteenth Century.* London: Methuen, 1970.

**[475] Sawada, S., "Technological Change in Japanese Agriculture: A Long-Term Analysis", Chapter 5 in *Agriculture and Economic Growth: Japan's Experience,* ed. by K. Ohkawa, B. Johnson and H. Kaneda. Tokyo: University of Tokyo Press, 1970.

[476] Scherer, F. M., "Research and Development Resource Allocation under Rivalry", *Quarterly Journal of Economics,* **81 (1967), 359–394.

**[477] Scherer, F. M., *Indistrial Market Structure and Economic Performance,* 2nd edn. Chicago: Rand McNally, 1980.

[478] Scherer, F. M., "Demand-Pull and Technological Invention: Schmookler Revisited", *Journal of Industrial Economics,* **30 (1982), 225–237.

**[479] Schiff, E., "Factor Substitution and the Composition of Input", in *Output, Input and Productivity Measurement,* Studies in Income and Wealth, Vol. 25, National Bureau of Economic Research, Princeton University, 1961.

[480] Schmookler, J., "The changing Efficiency of the American Economy", *Review of Economic and Statistics,* **34 214–232.

**[481] Schmookler, J., *Invention and Economic Growth.* Cambridge, Mass.: Harvard University Press, 1966.

**[482] Schon, D. A., *Beyond the Stable State.* New York: Random House, 1971.

[483] Schultz, T. W., *Transforming Traditional Agriculture.* New Haven: Yale University Press, 1964.

[484] Schultz, T. W., "Institutions and the Rising Economic Value of Man", *American Journal of Agricultural Economics,* **50 (1968), 1113–1122.

**[485] Schumpeter, J., *The Theory of Economic Development*. Cambridge, Mass.: Harvard University Press, 1934 (reprinted 1974, Oxford University Press).

[486] Schumpeter, J., "Alfred Marshall's Principles: A Semi-Centennial Appraisal", *American Economic Review*, **31 (2) (1941). Reprinted in J. Schumpeter, *Ten Great Economists*. London: Oxford University Press, 1951.

**[487] Schumpeter, J., *Business Cycles*, Vols. I and II. New York: McGraw-Hill, 1939.

**[488] Schutjer, W. and M. Van der Veen, "Economic Constraints on Agricultural Technology Adoption in Developing Countries", Occasional Paper No. 5, U.S. Agency for International Development, Washington, D.C., 1977.

[489] Shah, A. and M. Desai, "Growth Cycles with Induced Technical change", *Economic Journal*, **91 (1981), 1006–1010.

[490] Shakya, P. B. and J. C. Flinn, "Adoption of Modern Varieties and Fertilizer Use on Rice in the Eastern Taroi of Nepal", *Journal of Agricultural Economics*, **36 (1985), 409–419.

[491] Sharif, M. N. and A. K. Haq, "Determinant for Forecasting Technological Substitution", *Technological Forecasting and Social Change*, **13 (1979), 59–81.

[492] Sharif, M. N. and C. Kabir, "System Dynamics Modelling for Forecasting Multilevel Technological Sustitution", *Technological Forecasting and Social Change*, **9 (1976), 89–112.

[493] Sharif, M. N. and K. Ramanathan, "Binomial Innovation Diffusion Models with Dynamic Potential Adopter Population", *Technological Forecasting and Social Change*, **20 (1981), 63–87.

[494] Sharif, M. N. and K. Ramanathan, "Polynomial Innovation Diffusion Models", *Technological Forecasting and Social Change*, **21 (1982), 301–323.

[495] Shen, T. Y., "Innovation, Diffusion and Productivity Changes", *Review of Economics and Statistics*, **43 (1961), 175–181.

[496] Shen, T. Y., "Technology Diffusion, Substitution and X-efficiency", *Econometrica*, **41 (1973), 263–284.

[497] Sherwin, C. W. and R. S. Isenson, "Project Hindsight: A Defense Department Study of the Utility of Research", *Science*, **156 (1967), 1571–1577.

[498] Simon, H. A., *Administrative Behavior*, 2nd edn. New York: Free Press, 1965.

[499] Skiadas, C., "Two Generalized Rational models for Forecasting Innovation Diffusion", *Technological Forecasting and Social Change*," **27 (1985), 39–61.

[500] Skott, P., "Technological Advances with Depletion of Innovation Possibilities: A Comment and Some Extension", *Economic Journal*, **91 (364) (1981), 977–987.

**[501] Smiles, S., *Selections from Lives of the Engineers*, ed. and with Introduction by Thomas P. Hughes. Cambridge, Mass.: MIT Press, 1966.

[502] Smith, V. Kerry, "The Implications of Regulation for Induced Technical Change", *Bell Journal of Economics and Management Science*, **5 (1974), 623–632.

[503] Smith, V. K., "The Ames–Rosenberg Hypothesis and the Role of Natural Resources in the Production Technology", *Explorations in Economic History*, **15 (1978), 257–268.

[504] Soete, L. and R. Turner, "Technology Diffusion and the Rate of Technical Change", *Economic Journal*, **94 (1984), 612–623.

[505] Solow, R. M., "A Contribution to the Theory of Economic Growth", *Quarterly Journal of Economics,* **70 (1956), 65–94.
[506] Solow, R. M., "Technical Change and the Aggregate Production Function", *Review of Economics and Statistics,* **39 (1957), 312–320.
[507] Solow, R. M., "Some Recent Developments in the Theory of Production", in *The Theory and Empirical Analysis of Production,* ed. by Murray Brown. *Studies in Income and Wealth,* Vol. 31. New York: National Bureau of Economic Research, 1967.
**[508] Solow, R. and P. Temin, ""The Inputs for Growth", Chapter 3 in *The Economics of the Industrial Revolution,* ed. by J. Mokyr. London, Boston: George Allen & Unwin, 1985.
[509] Spence, A., "Investment Strategy and Growth in a New Market", *Bell Journal of Economics,* **10 (1979), 1–19.
[510] Spence, A., "The Learning Curve and Competition", *Bell Journal of Economics,* **12 (1981), 49–70.
[511] Staub, W. J. and M. G. Blase, "Induced Technological Change in Developing Agricultures: Implications for Income Distribution and Agricultural Development", *Journal of Developing Areas,* **8** (1974), 581–596.
[512] Stauffer, E. and M. Blase, "Institutional Disequilibria in the Development Process", *Economic Development and Cultural Change,* **22 (1974), 265–278.
[513] Steedman, I., "On the 'Impossibility' of Hicks-Neutral Technical Change", *Economic Journal,* **95 (1985), 746–758.
[514] Stevenson, R., "Measuring Technological Bias", *American Economic Review,* **70 (1980), 162–173.
**[515] Stewart, F. and J. James, *The Economics of New Technology in the Developing Countries.* London: Frances Pinter, 1982.
[516] Stigler, G., "The Economics of Information", *Journal of Political Economy,* **69** (1961), 213–255.
[517] Stigler, G., "Economic Problems in Measuring Changes in Productivity", in *Output, Input and Productivity Measurement,* Studies of Income and Wealth, Vol. 25, National Bureau of Economic Research, Princeton University, 1961.
[518] Stone, R. and D. Rowe, "The Market Demand for Durable Goods", *Econometrica,* **25** (1957), 423–443.
**[519] Stoneman, P., *Technology Diffusion and the Computer Revolution: The UK Experience.* Cambridge: Cambridge University Press, 1976.
[520] Stoneman, P., "The Rate of Imitation, Learning and Profitability", *Economic Letters,* **6 (1980), 179–183.
[521] Stoneman, P., "Intra-firm Diffusion, Bayesian Learning and Profitability", *Economic Journal,* **91 (1981), 375–388.
*[522] Stoneman, P., *The Econmoic Analysis of Technological Change.* London: Oxford University Press, 1983.
**[523] Stoneman, P., "Technological Diffusion: The Viewpoint of Economic Theory", *Warwick Economic Research Papers,* No. 270, Department of Economics, University of Warwick, Coventry, England, 1985.
[524] Stoneman, P. and P. A. David, "Adoption Subsidies vs. Information Provision as Instruments of Technology Policy", *Economic Journal, Supplement,* **96 (1986), 142S–150S.
[525] Stoneman, P. and N. J. Ireland, "The Role of Supply Factors in the Diffusion of New Process Technology", *Economic Journal, Supplement,* **93 (1983), 65S–77S.
**[526] Stoneman, P. and W. Ochoro, "A Means-Variance Approach to the Theory

of Intrafirm Diffusion", Chapter 2 in *The Economics of Technological Progress,* ed. by T. Puu and S. Wibe. London: Macmillan & Co., 1980.
**[527] Summers, G. F., ed., *Technology and Social Change in Rural Areas.* Boulder, Colo.: Westview Press, 1983.
[528] Sumrall, J. B., "Diffusion of the Basic Oxygen Furnace in the U.S. Steel Industry", *Journal of Industrial Economics,* **30 (1982), 421–437.
[529] Swann, P. L., "The International Diffusion of an Innovation", *Journal of Industrial Economics,* **22 (1973), 61–69.
[530] Sweeney, G., "Adoption of Cost-Saving Innovations by a Regulated Firm", *American Economic Review,* **71 (1981), 437–447.
[531] Takayama, A., "On Biased Technological Progress", *American Economic Review,* **64 (1974), 631–639.
**[532] Tarde, G., *The Laws of Imitation.* Translation, E. C. Parsons. New York: Holt, 1903.
[533] Teece, D. J., *The Multinational Corporation and the Resource Cost of International Technology Transfer.* Cambridge, Mass.: Ballinger, 1976.
[534] Teece, D. J., "Technology Transfer by Multinational Firms: The Resource Cost of Transferring Technology Know-how", *Economic Journal,* **87** (1977), 242–261.
**[535] Temin, P., "A New Look at Hunter's Hypothesis about the Antebellum Iron Industry", *American Economic Review, Papers and Proceedings* (1964), 344–351. Reprinted in *The Economics of Technological Change,* ed. by N. Rosenberg. London: Penguin, 1971.
**[536] Temin, P., *Iron and Steel in Nineteenth-Century America.* Cambridge, Mass.: MIT Press, 1964.
[537] Temin, P., "Labor Scarcity and the Problem of American Industrial Efficiency in the 1850's", *Journal of Economic History,* **26** (1966), 361–379.
[538] Temin, P., "Labor Scarcity in America", *Journal of Interdisciplinary History,* **1 (1971), 251–264.
**[539] Thirtle, C. G., "Induced Innovation in U.S. Agriculture", *Ph.D. dissertation,* Columbia University, 1982.
[540] Thirtle, C. G., "Testing the Induced Innovation Hypothesis: United States Agriculture, 1939–78", *Journal of Agricultural Economics,* **36 (1985), 1–14.
[541] Thirtle, C. G., "Accounting for Increasing Land-Labour Ratios in Developed Country Agricultures", *Journal of Agricultural Economics,* **36 (1985), 161–169.
**[542] Thirtle, C. G., "The Microeconomic Approach to Induced Innovation: A Reformulation of the Hayami and Ruttan Model", *The Manchester School,* (1985), 263–279.
[543] Thirtle, C. G., "Technological Change and the Productivity Slowdown in Field Crops: United States, 1939–78", *Southern Journal of Agricultural Economics,* **17 (2) (1985), 33–42.
**[544] Tobin, J. and W. Brainard, "Asset Markets and the Cost of Capital", Cowles Foundation Discussion Paper No. 427, Yale University, New Haven, March 1976.
[545] Uselding, P., "Technical Progress at the Springfield Armory, 1820–50", *Explorations in Economic History,* **9 (1972), 291–316.
[546] Uselding, P. and B. Juba, "Biased Technical Progress in American Manufacturing, 1839–1899", *Explorations in Economic History,* **11 (1973), 55–72.
**[547] Usher, A. P., *A History of Mechanical Inventions,* rev. ed. Cambridge, Mass.: Harvard University Press, 1954.

[548] Uzawa, H., "Neutral Inventions and the Stability of Growth Equilibrium", *Review of Economic Studies,* **28 (1961), 117–124.

[549] Uzawa, H., "Optimal Technical Change in an Aggregative Model of Economic Growth", *International Economic Review,* **6** (1965), 18–31.

**[550] van Duijn, J. J., *The Long Wave in Economic Life.* London, Boston: George Allen & Unwin, 1983.

[551] Vernon, R., "International Investment and International Trade in the Product Cycle", *Quarterly Journal of Economics,* **80 (1966), 190–207.

[552] Vernon, R., ed., *The Technology Factor in International Trade.* New York: Columbia University Press for the National Bureau of Economic Research, 1970.

[553] Vernon, R., "The Product Cycle Hypothesis in a New International Environment", *Oxford Bulletin of Economics and Statistics,* **41** (1979), 255–267.

**[554] Von Tunzleman, G., *Steam Power and British Industrialization to 1860.* Oxford: Clarendon Press, 1978.

[555] von Weizsacker, C. C., "Tentative Notes on a Two Sector Model with Induced Technical Progress", *Review of Economic Studies,* **33 (1966), 245–251.

**[556] Wade, W., *Institutional Determinants of Technical change and Agricultural Productivity Growth: Denmark, France and Great Britain, 1880–1965.* New York: Arno Press, 1981.

**[557] Wan, H. Y., *Economic Growth.* New York: Harcourt Brace and Jovanovich, 1971.

[558] Weaver, R. D., "Multiple Input, Multiple Output Production Choices and Technological in the U.S. Wheat Region", *American Journal of Agricultural Economics,* **65 (1983), 45–56.

**[559] Weber, A., "Productivity of German Agriculture: 1950–70", Staff Paper 73-1, Department of Agricultural and Applied Economics, University of Minnesota, St Paul, 1973.

[560] Wharton, C., "The Green Revolution: Cornucopia or Pandora's Box?" *Foreign Affairs,* **47** (1969), 464–476.

**[561] White, L., Jr, *Machine Ex Deo: Essays in the Dynamism of Western Culture.* Cambridge, Mass.: MIT Press, 1968.

**[562] Whitehead, A. N., *Science and the Modern World.* New York: Macmillan Co., 1925.

**[563] Wilkening, E. A., *Acceptance of Improved Farm Practices in Three Coastal Plain Countries,* Bulletin 98, North Carolina Agricultural Experiment Station, Raleigh, 1952.

**[564] Williams, B. R., *Science and Technology in Economic Growth.* London: Macmillan & Co., 1973.

[565] Williams, J. H., "The Theory of International Trade Reconsidered", *Economic Journal,* **39** (1929), 195–209.

[566] Williams, S. K. T., "Source of Information on Improved Farming Practices in Some Selected Areas of Western Nigeria", *Bulletin of Rural Economics and Sociology,* **4 (1969), 30–51.

[567] Williamson, J. G., "Optimal Replacement of Capital Goods: The Early New England and British Textile Firm", *Journal of Political Economy,* **79 (1971), 1320–1334.

[568] Williamson, J. G., "Technology, Growth and History", *Journal of Political Economy,* Part I, **8 (1976), 809–820.

[569] Willis, J., "Technical Change in the U.S. Primary Metals Industry", *Journal of Econometrics,* **10 (1979), 85–98.

[570] Wilton, P. C. and F. E. Pessemier, "Forecasting the Ultimate Acceptance of an Innovation: The Effects of Information", *Journal of Consumer Research,* **8 (1981), 162–171.

[571] Winter, S. G., "Schumpeterian Competition in Alternative Technological Regimes", *Journal of Economic Behavior and Organization,* **5 (1984), 287–320.

[572] Winter, S. G., "Economic 'Natural Selection' and the Theory of the Firm", *Yale Economic Essays,* **4 (1964), 225–272.

**[573] Winter, S. G., "Optimization and Evolution", in *Adaptive Economic Models,* ed. by R. Day and T. Groves. New York: Academic Press, 1975.

**[574] Woodland, A., "Modelling the Production Sector of an Economy: A Selective Survey and Analysis", Discussion Paper No. 76–21, Department of Economics, University of British Columbia, Vancouver, 1976.

**[575] Woodward, T., "A Factor Augmenting Approach for Studying Capital Measurement, Obsolescence, and the Recent Productivity Slowdown", Chapter 5 in *Developments in Econometric Analysis of Productivity: Measurement and Modelling Issues,* ed. by A. Dogramaci. The Hague: Kluwer Nijhoff, 1983.

*[576] Wyatt, G., *The Economics of Invention: A Study of the Determinants of Inventive Activity*. Brighton, England: Harvester Press, 1986.

[577] Yamada, S. and V. Ruttan, "International Companions of Productivity in Agriculture", in *New Developments in Productivity Measurement and Analysis,* ed. by J. W. Kendrick and B. N. Vaccara, NBER Studies in *Income and Wealth,* **44 (1980), 509–594. Chicago: University of Chicago Press, 1980.

[578] Yapa, L. S. and R. C. Mayfield, "Nonadoption of Innovation: Evidence from Discriminant Analysis", *Economic Geography,* **54 (1978), 145–156.

**[579] Yeomans, K. A., *Statistics for the Social Scientist,* Vol. 2: *Applied Statistics.* Harmondsworth, Middlesex: Penguin, 1968.

**[580] Yeung, P. and T. Roe, "A CES Test of Induced Technical Change—Japan", Chapter 8 in *Induced Innovation,* ed. by H. Binswanger and V. Ruttan. Baltimore: Johns Hopkins University Press, 1978.

[581] Zind, R. G., "A Note on the Measurement of Technological Bias in the U.S. Economy", *Review of Economics and Statistics,* **61 (1979), 301–304.

Index